INSTRUCTOR'S MANUAL TO
ACCOMPANY

OPERATIONAL
ORGANIC CHEMISTRY

INSTRUCTOR'S MANUAL TO
ACCOMPANY

OPERATIONAL
ORGANIC CHEMISTRY

A Laboratory Course

JOHN W. LEHMAN
Lake Superior State College

Allyn and Bacon, Inc.   *Boston*   *London*   *Sydney*   *Toronto*

ISBN 0-205-07147-3

Printed in the United States of America.

10 9 8 7 6 5 4 3 2   85 84 83 82 81

Preface

This manual is designed to assist the laboratory instructor by providing lists of needed chemicals and supplies, answers to questions and prelab assignments, and other helpful information about the experiments. The organization follows that in the text, with the Part I and Part II experiments listed by number, followed by material on Parts III and IV.

To help you design a laboratory course that is compatible with your lecture assignments and laboratory facilities, a list of textbook correlations and a list of the operations, by experiment, follow this preface. Although it is recommended that the first five experiments be performed in sequence, the order in which the rest are assigned can be quite flexible, depending on the students' preparation and abilities. Most of the Part II experiments are more demanding than those in Part I and should be assigned later in the course, but this is not invariably the case. Added flexibility is made possible by the 19 Minilabs and the numerous experimental variations described at the end of each experiment.

A general list of suggested laboratory equipment and supplies is provided below; special supplies required for certain experiments are listed under the experiments themselves. The operations section of the textbook (Part V) also lists equipment and supplies needed for performing each of the operations.

The experiment times are based on rough estimates of the number of 3- or 4-hour periods needed by most students to complete the experiments. If a range is given, the lower number applies to 4-hour periods. A plus sign (+) after the time generally means that the student must return to the laboratory briefly after the preparatory part is completed to gather analytical data, etc.

Quantities of chemicals and supplies listed under the experiments are based on ten students, with an allowance (usually about 30%) for waste and repetitions. In the Part I and II chemical lists, the quantities of liquids are given by both mass and volume, since suppliers may sell them either way. The volume of an aqueous solution is followed by the mass or volume of the solute, in parentheses. Quantities of all non-aqueous solvents, <u>including</u> those required for preparing solutions, are also listed. Directions for preparing any non-aqueous solutions or special reagents follow the chemical lists. For Part III (qualitative analysis) the volumes of non-aqueous solvents used in preparing solutions are given in the solution preparation section, but not in the chemical lists. Suppliers and catalog numbers are provided for the less common chemicals; alternate sources are available for virtually all of them. An effort has been made to keep costs down by using inexpensive chemicals or small amounts of the more expensive ones.

Answers are provided for all of the Topics for Report but not for the Library Topics; information on most of the latter can be found in sources listed in the Bibliography. Topics 1 and 2 should generally be assigned with each experiment since they often refer directly to observations or data from the experimental work; other topics can be assigned at your discretion.

Since many institutions cannot provide expensive instruments for routine use by undergraduates, a selection of spectra from the experiments requiring spectral interpretation is included in the back of this manual. They may be reproduced as necessary and distributed to students after the corresponding experiment has been completed.

The author welcomes all suggestions from users of Operational Organic Chemistry for improvements in the techniques, experimental procedures, or anything else that might make the book and this manual more useful.

John W. Lehman
Department of Chemistry
Lake Superior State College

Textbook Correlations

The first five experiments are not included in these correlations because their primary purpose is to teach basic lab
techniques, but they could be correlated with lecture material on carboxylic acids, esters, pharmaceuticals, etc.
if desired. I have tried to select the textbook chapters
bearing the most obvious relationship to the experimental
material, but in some cases there may be other chapters
that would correlate equally well. The third edition of
Morrison & Boyd is used here because the contents of the
fourth were not available at the time of publication of
this manual: I will be glad to provide correlations with
the fourth edition, on request, when it is released. The
following textbooks for the one-year course in organic
chemistry are included in the correlations:

M&B: R.T. Morrison and R.M. Boyd, Organic Chemistry, 3rd
    ed. Boston: Allyn & Bacon, 1973

Sol: T.W.G. Solomons, Organic Chemistry, 2nd ed. New York,
    Wiley, 1980

S&H: A. Streitwieser, Jr. and C.H. Heathcock, Introduction
    to Organic Chemistry. New York: Macmillan, 1976

Fes: R.J. Fessenden and J.S. Fessenden, Organic Chemistry.
    Boston: Willard Grant, 1979

R&C: J.D. Roberts and M.C. Caserio, Basic Principles of
    Organic Chemistry, 2nd ed. Menlo Park, CA: Benjamin,
    1977

PHCH: S.H. Pine, J.B. Hendrickson, D.J. Cram and G.S.
    Hammond, Organic Chemistry, 4th ed. New York: McGraw-
    Hill, 1980

K&V: D.S. Kemp and F. Vellaccio, Organic Chemistry. New
    York: Worth, 1980

Ter: A.L. Ternay, Contemporary Organic Chemistry. Phila-
    delphia: Saunders, 1976

Exp. No.	M&B	Sol	Textbook Chapter(s) S&H	Fes	R&C	PHCH	K&V	Ter
6	1	2	2	2	2	2	1	2
7	3	3	4	3	4	2	14	3
8	5	6	12	7	8	10	7 17	6
9	3	3	4	3	4	2	14	3
10	6 7	7 8	7 12	4 9	4 10	4 11	5 16	8
11	8	9	13	9	10	11	18	9
12	9	3 8	23	9	12 13	4 11	13 32	7 8
13	10	11	21 36	10	21	13	20 32	14
14	12	12	5	6 10	4 26	18	14 32	15
15	13	10	22 36	20	9 30	11 12	19	28
16	14	14	6	5 7	14	9	3 8	5
17	15	14 15	9 15	6 7	14	7	8 32	10
18	16	15	11	7	15 30	10 11	8 9	10
19	18	17	17	12	18	8	10 11	18
20	19	16	15	11	16	7	9	16
21	20	17	18 27	13	24	8	11	18
22	13 24	13 15	14 33	7 8	9 26	5	12 21	22 28
23	34	19	25	17	20	15	27	25
24	36	22	28	18	25	16	28	24
25	4	8	7	4	4 19	4	5	4
26	8	10	20 23	9	13	12	19 30	12 13
27	11 19	12	21 31	10	22	13	22	15
28	11	12	21 29	10	22	13	20	15
29	12 13	12 13	14 30	8 10	9 26	5 13	12 22	15 28
30	14	5	8	5	8	6 9	7	5
31	16	15	11	7	15	10	8	10
32	17	15	11 33	7	15	9	3	11
33	19	15	15	11	16	7	9	16
34	19 28	16	24	14	17 26	19	26 31	16 17
35	13 21	13 16	10 15	8 14	9	5 7	12 24	17 29
36	23	18	32 36	15	23 28	13	21	21
37	23	18	27 31	13 15	23	8	34	23
38	25	14	30	10	14	6 13	20	21
39	26	20	27	14	17	7 9	24	20
40	21 27	16 20	15 26	11 14	17	8 12	23 24	17 19
41	30	11	34	16	22 26	13	22	23
42	31	11	35	16	30	13	34	15
43	32	8	12	9	29	20	29 38	8
44	33	21	18	19	18	17	10	27

Use of the Operations by Experiment

The Elementary Operations, OP-1 through OP-6, are omitted.
Some obvious or trivial applications of certain operations
are not listed; for example, heating (OP-7) is used in all
reflux operations (OP-7a) and boiling point measurements
(OP-29) are made during most distillations. Many operations,
particularly OP's 7b, 12a, 13a, 13c, 20, 21, 23a, 23b, 23c,
25a, 26a, 28, 29, 30, 32, 33, 34, and 35 are also used in
Part III and in the Experimental Variations.

Operation		Experiments
7	Heating	3 4 18 23 26 29 30 37 39 42 43 44
7a	Refluxing	3 5 10 12 16 17 18 19 21 26 27 32 34 37 39 40 41 43
7b	Semimicro Refluxing	Part III only
7c	Temperature Monitoring	4 31
8	Cooling	13 21 35 36 37
9	Mixing	12 13 21 28 31 32 34 43
10	Addition	10 11 17 18 21 27 39 41
10a	Semimicro Addition	14
11	Gravity Filtration	1 4 5 14 23 34 37 44
12	Vacuum Filtration	1 3 4 10 13 15 17 18 23 26 27 29 34 35 36 37 40 41 42 43
12a	Semi. Vac. Filtration	14 17 20
13	Extraction	1 11 12 21 22 29 31 32 35 39 44
13a	Semimicro Extraction	24
13b	Separation of Liquids	12 16 17 18 27 29 32 39
13c	Salting Out	11
13d	Solid Extraction	15 35
14	Evaporation	1 11 12 14 15 17 21 22 24 25 27 31 32 35 39 40 44
15	Codistillation	11 16
15a	Steam Distillation	22
16	Column Chromatography	15 25
17	Thin-layer Chrom.	20
18	Paper Chromatography	24
19	Washing Liquids	5 8 15 16 17 18 27 28 31 32 35
20	Drying Liquids	5 8 12 15 16 17 18 21 22 27 28 32 35 37 39
21	Drying Solids	1 2 3 4 11 13 14 17 18 23 26 29 34 35 36 37 40 41 42 43
22	Drying Gases	24
22a	Excluding Moisture	17 39 40
22b	Trapping Gases	14 16 21 27 37
23	Recrystallization	2 3 17 26 29 34 35 37 40 42
23a	Semi. Recrystallization	14 20
23b	Recryst./Mixed Solvents	4 20 35
23c	Choosing Solvents	Part III only
24	Sublimation	18
24a	Vacuum Sublimation	41
25	Simple Distillation	5 11 16 28 31
25a	Semi. Distillation	7 8 12 27 32 43
25b	Dist. of Solids	35
25c	Water Separation	19 39
26	Vacuum Distillation	21 39
26a	Semi. Vacuum Dist.	21 35
27	Fractional Distillation	8 9
28	Melting Point	2 3 10 14 17 18 20 23 26 27 29 34 35 37 40 41 42
29	Boiling Point	7
29a	Micro Boiling Point	7
30	Refractive Index	7

31	Optical Rotation	23 25
32	Gas Chromatography	8 9 21 22 28 44
33	Infrared Spectrometry	11 22 25 29 43
34	NMR Spectrometry	13 35 40
35	Vis-UV Spectrometry	15
35a	Colorimetric Analysis	38

Laboratory Equipment and Supplies

Each student should have access to an organic glassware kit
(preferably 19/22) or the equivalent, including the following: round-bottom flasks (25, 50, 100, 250, and 500 ml),
Liebig-West condenser, distilling column, vacuum adapter,
three-way connecting tube, Claisen connecting tube, thermometer adapter, separatory/addition funnel, and glass (or
plastic) stopper. These can be provided in student lockers
or checked out of the stockroom when needed. Students
should also be provided with (or required to purchase)
safety glasses and "rubber" gloves; a towel, sponge, and
lab apron are also desirable.

Each lab station should be supplied with at least 2 ring-
stands (3 preferred), 3 utility clamps, 1 condenser clamp,
2 ring supports, a steam bath, and a burner (with flame
spreader). Students should have access to one or more
balances, drying ovens, gas chromatographs, infrared spectrophotometers, and refractometers, if possible. A flame-
less heat source, such as a heating mantle (we find the
250-ml Briskeat beaker heater quite satisfactory) is high-
ly desirable, and magnetic stirrers (with stirring bars)
are convenient for many experiments, though they can be
done without. Instruments that are used less frequently
and are therefore optional include polarimeters, NMR spectrometers, ultraviolet-visible spectrophotometers, color-
imeters, and pH meters. (Note that all of the spectral
determinations can be dry-labbed, if necessary.) Melting-
point instruments will reduce the time required for m.p.
determinations considerably if they are available.

Community supplies that should be readily available in the
lab or stockroom include the following: soap, brushes,
pipe cleaners, wash acetone, decolorizing carbon, Drierite
(or calcium chloride), filter-aid, stopcock lubricant,
glass rod, glass tubing, melting-point capillaries, column
packing, boiling chips, boiling sticks (wooden applicators),
rubber bands or spring clamps, rubber stoppers, corks, cork
borers, triangular files, and scissors. Special equipment
like chromatography columns (listed in this manual under
the experiment or in the textbook under the operation) can
be issued from the stockroom as required. A suggested
locker list of frequently used supplies follows:

2 beakers, 50 ml
2 beakers, 150 ml
beaker, 250 ml
beaker, 600 ml
beaker, 1000 ml
bottle, screw-cap, 125 ml
bottle, squat form, screw-
 cap, 250 ml (8 oz)
bottle, wide mouth, 500 ml
bulb, pipetting
burner lighter
cork ring, 60 mm
cork ring, 120 mm
cylinder, graduated, 25 ml
cylinder, graduated, 100 ml
drying tube, 100 mm
evaporating dish, 100 mm
2 flasks, Erlenmeyer, 50 ml
2 flasks, Erlenmeyer,
 125 ml
2 flasks, Erlenmeyer,
 250 ml
flask, Erlenmeyer, 500 ml
flask, filter, 250 ml
funnel, Buchner, 55 mm
funnel, Hirsch, Coors 4/0
funnel, powder, 80 mm
funnel, short-stem, 65 mm
litmus paper, blue
litmus paper, red

3 medicine droppers
pinch clamp
pH paper, wide range
pipet, graduated, 1 ml
pipet, graduated, 10 ml
2 pipets, Pasteur, with bulbs
ruler, metric
2 screw clamps
spatula, micro
spatula, wooden-handled
6 test tubes, 13 x 100 mm
4 test tubes, 18 x 150 mm
2 test tubes, 25 x 200 mm
test tube holder
test tube rack
thermometer, 360°C
Thiele tube
tongs, steel
3 tubes, rubber, 3' x ¼"
2 tubes, rubber, 2' x 3/16",
 pressure
tubing connector, T-shape
3 vials, 2 dram
3 vials, 5 dram
wash bottle, polyethylene,
 125 ml
3 watch glasses, 50-110 mm
wire gauze, asbestos center
wire gauze, plain

PART I

MAIN SEQUENCE EXPERIMENTS

PRELIMINARIES

Time: ½-1 period

Supplies and chemicals per 10 students:

bottles, wide mouth, 250 ml (Pyrex, preferably)	10
bunsen burners	10
cork borer, #6	2
files, triangular	2
flame spreaders	10
glass rod, soft glass, 5-6 mm, 4' lengths	3
glass tubing, Pyrex, 8 mm O.D., 4' lengths	3
glass tubing, soft glass, 4-5 mm O.D., 4' lengths	1½
glycerin, in dropper bottles	2 btls
methylene chloride (to test boiling tubes)	25 g (20 ml)
rubber stoppers, solid, #6 (to fit 250 ml filter flask)	15
rubber stoppers, 2-hole (to fit 250 ml bottles)	10
rulers, metric	2
tape, plastic	1 roll

EXPERIMENT 1

Time: 1 period

Chemicals per 10 students:

acetanilide	20 g
aspirin (acetylsalicylic acid)	15 g
calcium chloride or Drierite (for dessicator jars)	750 g
3M hydrochloric acid	260 ml (65 ml conc. HCl)
methylene chloride (dichloromethane)	850 g (640 ml)
1M sodium hydroxide	650 ml (26 g NaOH)
starch	5 g

Comments: "Panacetin" can be prepared by thoroughly mixing acetanilide, aspirin, and starch in the proportions 20 g: 15 g: 5 g, respectively (or different proportions can be used). The analysis of this fictional analgesic drug is divided into two parts so that the students will have time to read and assimilate all of the assigned operations and appendixes. Students should save the dessicant (in tightly capped jars) for use in subsequent experiments. The most common student errors are insufficient mixing during the extraction and insufficient drying of the aspirin; if possible, the aspirin should be dried in a 90° oven or allowed to dry overnight or longer in a dessicator jar.

Topics for Report:

1. To decrease the solubility of the aspirin and increase the recovery. It was hot because of the exothermic reaction $HCl + NaOH \rightarrow NaCl + H_2O + heat$.

2. Some aspirin would remain in the organic layer. This aspirin would be isolated with the acetanilide after evaporation; the reported percentage of acetanilide would be too high and that of aspirin too low.

3. (a) Acetaminophen would be extracted from the methylene chloride layer by aq. NaOH, along with the aspirin. (b) Extract aspirin with aq. sodium bicarbonate, then extract acetaminophen with aq. sodium hydroxide.

4.

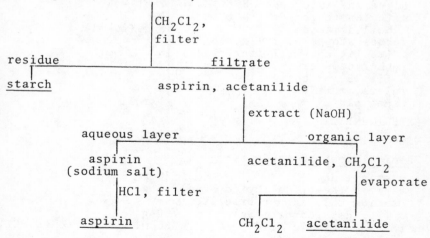

EXPERIMENT 2

Time: 1 period

Prelab calculation: volume of water = (g acetanilide recovered) x 100 ml/3.5 g.

2

  melting-point tubes, capillary, sealed, 90-100 mm   75
                     or
  glass tubing, m.p., capillary, 1.5 mm O.D.,
  1' lengths   25

Comments: Oiling may occur during recrystallization if
much less than the calculated volume of water is used. It
may be necessary to dry the acetanilide in an oven or over
a steam bath if students are to finish the experiment in
one period. Each student should be provided with about
2 12" lengths of capillary tubing for making his/her
melting-point tubes, or with 6-8 pre-sealed tubes. The
number of m.p. determinations (minimum of six per student)
can be reduced if desired, but the purpose is to give
students enough practice with melting point measurements
so that they will become adept at it.

Topics for Report:

1. (a) Losses during extraction, recrystallization,
transfers etc. (b) Mixture melting point should confirm
identity of adulterant as acetanilide.

2. Maximum recovery = g acetanilide - [(ml water) x
0.56 g/100 ml].

3. m-Aminophenol, since m.p. of a mixture of the unknown
with this compound is not lowered.

4. (a) Dispersion forces, ion-dipole interactions,
dipole-dipole interactions, hydrogen bonds. Polar
compounds tend to dissolve in polar solvents, non-polar
compounds in non-polar solvents. (b) Dispersion;
ion-dipole; hydrogen bonding.

EXPERIMENT 3

Time: 1½-2 periods

Prelab calculations: 5.51 g (4.69 ml) methyl salicylate

Chemicals per 10 students:

  methyl salicylate   70 g (60 ml)
  5M sodium hydroxide   650 ml (130 g NaOH)
  2M sulfuric acid   900 ml (100 ml conc.
                                 sulf. acid)

Comments: I encourage students to weigh liquid limiting
reactants, after measuring the approximate amounts by
volume, so that they can determine their percent yields
with some accuracy. If time or equipment is limited,
however, the methyl salicylate can be measured by volume
only. If methyl salicylate and the NaOH solution are
poured into the reaction mixture through the same funnel,
the precipipate that forms may plug up the funnel (it
can be dissolved with a little water). Since the vacuum
filtration operation includes directions to wash the
precipitate on the funnel with cold solvent, washing is

not generally specified in the procedure unless the solvent is different from that in the filtrate. Students may have to be reminded periodically of the necessity of washing, however.

Topics for Report:

1. (a) $5.00 \text{ g} - 90 \times \frac{0.25}{100} = \underline{4.78 \text{ g}} \ (\underline{95.6\%})$

   (b) $5.00 \text{ g} - 90 \times \frac{0.10}{100} = \underline{4.91 \text{ g}} \ (\underline{98.3\%})$

2. 50 ml NaOH x 5M = 250 mmol = $\underline{0.25 \text{ mol}}$ = $\underline{0.25 \text{ eq}}$
   70 ml H_2SO_4 x 2M = 140 mmol = $\underline{0.14 \text{ mol}}$ = $\underline{0.28 \text{ eq}}$, so the acid is 0.03 eq in excess.

3. $(H^+) = 10^{-14}/3.5 = 2.9 \times 10^{-15}$ M; $\frac{(A)}{(B)} = \frac{(H^+)}{K_2} = $
   $\frac{2.9 \times 10^{-15}}{3.6 \times 10^{-14}} = 0.081$

   $\%A = \frac{(A)}{(A) + (B)} \times 100 = \frac{0.081}{1.081} \times 100 = \underline{7.5\%}$

4. The sodium salt of methyl salicylate:

   (structure: benzene ring with O^-Na^+ and $COOCH_3$ substituents)

EXPERIMENT 4

Time: 1 period

Chemicals per 10 students:

acetic anhydride	140 g (130 ml)
ethanol (95% or absolute)	175 g (225 ml)
*1% ferric chloride	1 ml
3M hydrochloric acid	300 ml (75 ml conc. HCl)
**salicylic acid	0.2 g
5% sodium bicarbonate	800 ml (40 g $NaHCO_3$)
sulfuric acid, conc.	12 g (6.5 ml)

*stabilize with a drop or two of dilute HCl
**additional sal. acid may be needed for students who obtained less than 3 g from Experiment 3.

Comments: It is important to avoid overheating during the reaction, because the resulting sticky polymer makes it difficult to work up the reaction mixture. If the aspirin is heated too long or at too high a temperature during recrystallization, it may partly hydrolyze to salicylic acid; the ferric chloride test will then show the recrystallized product to be less "pure" than the crude aspirin! Another common mistake is using too much ethanol during recrystallization when the yield of aspirin is less than average; the excess solvent can be evaporated, but

4

aspirin will be partly hydrolyzed in the process. Pure
aspirin should yield no purple coloration in the ferric
chloride test. Students find E.V. 4 interesting since
aspirin that has stood on the shelf for some time will show
evidence of salicylic acid.

Topics for Report:

1. $\dfrac{5 \text{ g NaHCO}_3}{100 \text{ g solution}} \times \dfrac{1.034 \text{ g soln.}}{1 \text{ ml soln.}} \times 60 \text{ ml soln.} \times$

   $\dfrac{1 \text{ mol NaHCO}_3}{84 \text{ g NaHCO}_3} \times \dfrac{1 \text{ mol HCl}}{1 \text{ mol NaHCO}_3} \times \dfrac{1 \text{ lit soln.}}{3 \text{ mol HCl}} \times \dfrac{1000 \text{ ml}}{1 \text{ lit}} =$

   12.3 ml of 3M HCl

2.

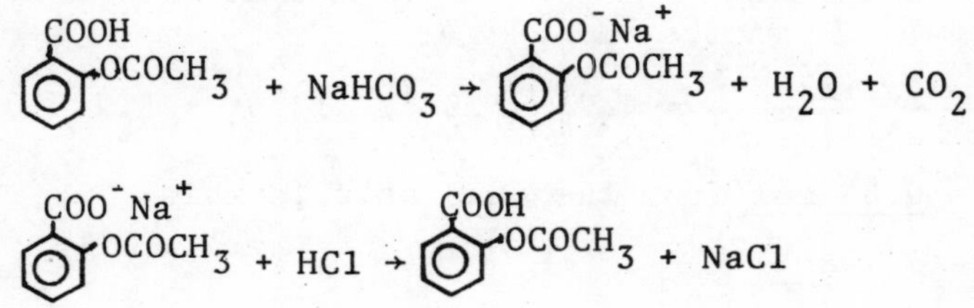

3. $\dfrac{0.325 \text{ g} \times 100}{180.2}$ = 0.180 mol aspirin (and acetic anhydride,
   salicyclic acid). Salicylic acid:

   $\dfrac{0.180 \text{ mol} \times 138 \text{ g} \times 1 \text{ lb} \times 90\cancel{c}}{1 \text{ mol} \quad 454 \text{ g} \quad 1 \text{ lb}}$ = 4.9¢

   acetic anhydride: $\dfrac{0.180 \text{ mol} \times 102 \text{ g} \times 1 \text{ lb} \times 27\cancel{c}}{1 \text{ mol} \quad 454 \text{ g} \quad 1 \text{ lb}}$ = 1.1¢

   total: 6.0¢

EXPERIMENT 5

Time: 1-1½ periods

Prelab exercise 2: Change 11.1 g (13.7 ml) of 1-butanol
to 13.1 g (16.2 ml) of isoamyl alcohol; change butyl
acetate at 120-126° to isoamyl acetate at 137-143°.

Chemicals per 10 students:

acetic acid, glacial	230 g (220 ml)
magnesium sulfate, anhydrous	25 g
3-methyl-1-butanol (isoamyl alcohol)	170 g (210 ml)
sodium bicarbonate, saturated aq. soln.	450 ml (40-45 g $NaHCO_3$)
sulfuric acid, conc.	25 g (14 ml)

5

Comments: From OP-19 and OP-20, students should estimate that about 30 ml of aqueous sodium bicarbonate (15 ml per portion) and 1 g of anhydrous magnesium sulfate are required for washing and drying the product. A volatile forerun can be expected during the distillation, and students should not begin collecting the product until the temperature reaches about 137°. The refractive index of isoamyl acetate (E.V. 1) is reported to be 1.4003 at 20°C. Very dilute solutions of isoamyl acetate are said to have an odor of pears, while the more concentrated material has a banana odor (E.V. 4).

Topics for Report:

1. (a) $\dfrac{(x)(x)}{(1-x)(1-x)} = 4.2$; $\dfrac{(x)}{(1-x)} = \sqrt{4.2}$;

    $x = \underline{0.67 \text{ mol}}$ BuOAc Yield $= \dfrac{0.67 \text{ mol}}{1 \text{ mol}} \times 100 = 67\%$

   (b) $\dfrac{(x)(x)}{(2-x)(1-x)} = 4.2$; $3.2x^2 - 12.6x + 8.4 = 0$,

    $x = \underline{0.85 \text{ mol}}$ (from the quadratic formula);

    yield $= \underline{85\%}$

2. (a) $H_2SO_4 + 2NaHCO_3 \rightarrow Na_2SO_4 + 2H_2O + 2CO_2$

    $HOAc + NaHCO_3 \rightarrow NaOAc + H_2O + CO_2$

   (b) The resulting salts are soluble in the aqueous layer, insoluble in the organic layer.

3. From the CRC Handbook:

	MW	bp	d
propionic acid (propanoic acid)	74.1	141°	0.993
isobutyl alcohol (2-methyl-1-propanol)	74.1	108°	0.802
isobutyl propionate	130.2	137°	0.869

Use 11.1 g (13.9 ml, 0.15 mol) of isobutyl alcohol and 22.2 g (22.4 ml, 0.30 mol) of propionic acid in the reaction. Collect the product at 132-138°.

EXPERIMENT 6

Time: 1 period

Chemicals per 10 students:

acetone	65 g (80 ml)
butanal (butyraldehyde)	10 g (12 ml)
1-butanol	10 g (12 ml)
2-butanone	10 g (12 ml)

n-butylamine	9 g (12 ml)
1-chloropropane	
(or 2-chloropropane)	12 g (14 ml)
chromium trioxide (chromic	
anhydride)	1½ g
copper wires, 0.5 mm O.D.,	
10 cm lengths	10
2,4-dinitrophenylhydrazine	4 g
*95% ethanol	200 g (260 ml)
pentane	9 g (15 ml)
"pentene" (Eastman Kodak T418,	
etc.)	10 g (15 ml)
1% potassium permanganate,	
aqueous	10 ml (0.1 g $KMnO_4$)
propanoic acid	12 g (12 ml)
sulfuric acid, conc.	40 g (22 ml)

*Includes 100 ml for preparing DNPH solution.

Solution preparations: 2,4-Dinitrophenylhydrazine reagent mix 4 g of 2,4-DNPH with 20 ml conc. sulfuric acid and add 30 ml of water slowly, with stirring, to dissolve the solid. Add 100 ml of 95% ethanol to the warm solution and filter if necessary. Chromic acid reagent (for 20 students) slowly pour a suspension of 3 g CrO_3 in 3 ml conc. sulfuric acid into 9 ml of water, with stirring.

Comments: Other appropriate low M.W. liquids can, in some cases, be substituted for those specified without changing the results significantly. For example, 2-chloropropane can be used in place of the more expensive 1-chloro compound, and cyclohexene can be substituted for the commercial pentene mixture. The butanal should be newly opened or recently distilled; otherwise it will give a faint test with blue litmus paper. If the experience with pipetting liquids is not considered necessary, the unknowns can be set out in dropper bottles and their volumes measured in drops for the solubility determinations.

Topics for Report:

1.

Compound	odor	solu-bility	d.	b.p.
butanal	sweaty-fruity, strong	sol.	0.817	76⁰
1-butanol	spiritous, mild	sol.	0.810	117⁰
2-butanone	"nail-polish remover", m-str.	sol.	0.805	80⁰
butylamine	ammoniacal, strong	misc.	0.741	78⁰
1-chloropropane	ethereal, strong	insol.	0.891	47⁰
pentane	"dry cleaning fluid", med.	insol.	0.626	36⁰
2-pentene	pungent, medium	insol.	0.648	36⁰
propanoic acid	very pungent (vinegar), str.	misc.	0.993	141⁰

Odor designations are very subjective - those given are typical student responses. Students should explain how classification test results prove the identity of each compound: e.g. the unknown that reacts with both DNPH and chromic anhydride must be butanal, etc. Butanal often appears insoluble, in our experience.

2.
(a) 1. pentane, pentene — dispersion forces only
2. 1-chloropropane — weak-dipole-dipole interaction
3. butanal, 2-butanone, butylamine — strong dipole-dipole interaction (weak hydrogen bonding in the case of butylamine)
4. 1-butanol — hydrogen-bonding
5. propanoic acid — multiple hydrogen bonding

(b) 1. pentane, pentene — no heteroatoms
2. butylamine — 1 nitrogen atom per molecule
3. butanal, 1-butanol, 2-butanone — 1 oxygen atom per molecule
4. 1-chloropropane — 1 chlorine atom per molecule
5. propanoic acid — 2 oxygen atoms per molecule

(c) <u>Insoluble</u> - pentane, pentene, 1-chloropropane — no hydrogen bonding with water
<u>Soluble</u> - 1-butanol, butanal, 2-butanone — 1 oxygen atom available for hydrogen bonding with water
<u>Miscible</u> - butylamine, butanoic acid — multiple hydrogen bonding with water (2 NH bonds on amine, 1 OH bond plus one oxygen atom on the acid)

3.
(a) A - aldehyde and ketone; B - ketone; C - alkene, aldehyde, alcohol; D - alkene, carboxylic acid.

(b)

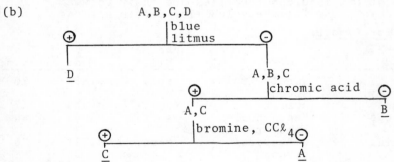

Compound A may decolorize bromine, but with evolution of HBr.

4. $CH_3CH_2CH_2CHO + ArNHNH_2 \rightarrow CH_3CH_2CH_2CH=NNHAr + H_2O$

$CH_3CH_2COCH_3 + ArNHNH_2 \rightarrow CH_3CH_2\underset{\underset{CH_3}{|}}{C}=NNHAr + H_2O$

8

$$3CH_3CH_2CH_2CH_2OH + 2H_2Cr_2O_7 + 6H_2SO_4 \rightarrow 3CH_3CH_2CH_2COOH +$$

$$2Cr_2(SO_4)_3 + 11\ H_2O$$

$$3CH_3CH_2CH_2CHO + H_2Cr_2O_7 + 3H_2SO_4 \rightarrow 3CH_3CH_2CH_2COOH +$$

$$Cr_2(SO_4)_3 + 4H_2O$$

$$3CH_3CH{=}CHCH_2CH_3 + 2KMnO_4 + 4H_2O \rightarrow 3CH_3\overset{OH}{\underset{|}{CH}}{-}\overset{OH}{\underset{|}{CH}}CH_2CH_3 +$$

$$2MnO_2 + 2KOH$$

$$(Ar = O_2N{-}\!\bigcirc\!{-})$$
$$NO_2$$

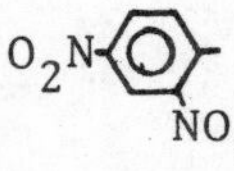

EXPERIMENT 7

Time: 1 period

Chemicals per 10 students:

  hydrocarbon unknowns (from Table 1)   90 g (130 ml)

Special apparatus: refractometers

Comments: Rather than setting out a single unknown labeled SRTC-491, students can be issued about 10 ml each of different hydrocarbons in numbered test tubes or vials. The hydrocarbons are available from Aldrich Chemical Co.: all but 2,3- and 2,4-dimethylpentane can be purchased for about $10 or less per 100 g (1980 prices). Students should discard a few ml of forerun and residue and use only the middle distillation fraction for determining the physical properties (particularly refractive index). Most unknowns are identified correctly with no difficulty; the most common error in our experience is misidentification of 2,2,4-trimethylpentane as 2,3-dimethylpentane.

Topics for Report:

1. Put a little of the liquid in water and see if it sinks or floats. Some students, remembering Experiment 6, may want to perform a Beilstein test using (for example) a copper scouring pad and a gas burner. Answers that would involve trying to measure the value of the boiling point, density, or refractive index in the kitchen are less acceptable because they lack simplicity and are subject to considerable error.

2. b.p. (corr) = $78.2^0 + 1.2 \times 10^{-4}(26\ torr)(351.3\ K) =$
   $79.3°C$
   n(corr) = $1.3780 + 8 \times 0.00045 = 1.3816$
   d = 3.346 g/5 ml = 0.669 g/ml

   The unknown is 2,4-dimethylpentane

3.

$$C-C-C-C-C-C-C \qquad \overset{\displaystyle C}{\overset{|}{C-C-C-C-C-C}} \qquad \overset{\displaystyle \quad C}{\overset{\quad |}{C-C-C-C-C-C}}$$

heptane   2-methylhexane   3-methylhexane

$$\overset{\displaystyle C-C}{\underset{|}{C-C-C-C-C}} \qquad \overset{\displaystyle C}{\underset{\underset{\displaystyle C}{|}}{\overset{|}{C-C-C-C-C}}} \qquad \overset{\displaystyle \quad C}{\underset{\underset{\displaystyle \quad C}{\quad |}}{\overset{\quad |}{C-C-C-C-C}}}$$

3-ethylpentane

               2,2-dimethylpentane   3,3-dimethylpentane

$$\overset{\displaystyle C \;\; C}{\underset{\underset{\displaystyle C}{|}}{\overset{| \;\; |}{C-C-C-C}}}$$

2,2,3-trimethylbutane

(3-methylhexane exists as 2 enantiomers)

4.

$$C-C-C-C \xrightarrow[\text{heat, pressure}]{AlCl_3} \overset{\displaystyle}{\underset{\underset{\displaystyle C}{|}}{C-C-C}} \quad \text{(isomerization)}$$

$$\overset{}{\underset{\underset{\displaystyle C}{|}}{C-C-C}} \xrightarrow[\text{heat}]{\text{catalyst,}} \overset{}{\underset{\underset{\displaystyle C}{|}}{C-C=C}} \quad \text{(dehydrogenation)}$$

$$\underset{\underset{\displaystyle C}{|}}{C-C-C} + \underset{\underset{\displaystyle C}{|}}{C-C=C} \xrightarrow{H^+} \overset{\displaystyle C}{\underset{\underset{\displaystyle C \;\; C}{| \;\; |}}{\overset{|}{C-C-C-C-C}}} \quad \text{(alkylation)}$$

EXPERIMENT 8

<u>Time</u>: 2 periods

<u>Prelab calculation</u>: $\dfrac{20 \text{ ml} \times 0.802 \text{ g/ml}}{102.2 \text{ g/mol}} \times \dfrac{82.4 \text{ g/mol}}{0.67 \text{ g/ml}} =$

     19.7 ml alkenes

<u>Chemicals per 10 students</u>

calcium chloride, anhydrous	15 g
4-methyl-2-pentanol (Aldrich 10,991-6)	300 g (240 ml)
2-methyl-2-pentene (Fluka 68460)*	2 g (3 ml)
5% sodium bicarbonate	150 ml (7.5 g NaHCO$_3$)
sodium chloride, satd. aq. soln.	150 ml (∿50 g NaCl)
sulfuric acid, conc.	45 g (25 ml)

<u>Special apparatus</u>: gas chromatographs(s)

<u>Comments</u>: Although the operation used in the <u>Reaction</u> step is not, strictly speaking, fractional distillation, it does

*Can also be obtained from Aldrich (Cat.#M6, 730-3).

give students experience with the techniques involved.
Most students collect 17-18 ml of alkene during this step.
Separation: About 10 ml of each wash solvent and 1 g of
calcium chloride should be used for washing and drying.
Analysis: We have used a 3% SE-30 stationary phase on
Aeropak #30 at 30^0 for the GLC separation; an SF-96/
Chromosorb column is also satisfactory. The order of
elution (on silicon oil) is 4-methyl-1-pentene, 4-methyl-
2-pentene, 2-methyl-1-pentene and 2-methyl-2-pentene.
4-methyl-2-pentene is ordinarily the major product (60-70%)
with 2-methyl-2-pentene constituting most of the balance.
E.V. 1: Using phosphoric acid as the catalyst results in a
longer reaction time but gives less tar formation. With
this acid, the product mixture usually contains a greater
proportion of the thermodynamic product, 2-methyl-2-pentene.

Topics for Report:

1. (a) 4-methyl-2-pentene; (b) 2-methyl-2-pentene

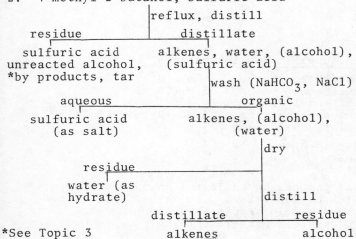

The Saytzev product, which is the most stable alkene that
can be formed without a hydride shift, predominates;
apparently the reaction conditions are not sufficient to
form the thermodynamically favored product in the highest
yield.

2. 4-methyl-2-butanol, sulfuric acid
                    |reflux, distill
    residue _____|___ distillate
    sulfuric acid   alkenes, water, (alcohol),
  unreacted alcohol,   (sulfuric acid)
  *by products, tar
                    |wash (NaHCO₃, NaCl)
      aqueous _____|____ organic
    sulfuric acid   alkenes, (alcohol),
      (as salt)   (water)
                        |dry
      residue _____|
    water (as
    hydrate)
                        |distill
      distillate _____|___ residue
*See Topic 3   alkenes   alcohol

11

Note: substances present only in trace quantities are set off in parentheses.

3.

$$\underset{\overset{|}{C}}{C-C-C-C-C} \oplus$$

$$+ \ C-C-C-\overset{\overset{OH}{|}}{C}-C \ \xrightarrow{H^+} \ C-C-C-\overset{\overset{|}{C}}{C}-O-\overset{\overset{|}{C}}{C}-C-\overset{\overset{|}{C}}{C}-C \quad \underline{A}$$

$$+ \ HSO_4^- \ \longrightarrow \ C-C-C-\overset{\overset{|}{C}}{C}-OSO_3H \quad \underline{B}$$

$$+ \ C-\overset{\overset{|}{C}}{C}-C=C-C \ \xrightarrow{-H^+} \ C-C-C=\overset{\overset{|}{C}}{C}-\overset{\overset{|}{C}}{C}-C-\overset{\overset{|}{C}}{C}-C \quad \underline{C}$$

(and other isomers)

Most of the by-products will be left behind in the residue of the first distillation; traces of B that distill over will be removed during the sodium bicarbonate washing, and A and C during the final distillation.

4.

$$\text{Volume water expected} = \frac{20 \text{ ml} \times 0.802 \text{ g/ml}}{102.2 \text{ g/mol}} \times$$

$$\frac{18.0 \text{ g/mol}}{1.00 \text{ g/ml}} = 2.83 \text{ ml}$$

$$\text{Ratio} = \frac{2.83}{19.7} \text{ (from prelab)} = 0.143 \text{ (about 1:7)}$$

5. (a) $C-\overset{\overset{|}{C}}{C}-\overset{\overset{|}{C}}{C}=C-C$; ; $C-\overset{\overset{|}{C}}{C}-C=C$

(b) $C-\overset{\overset{|}{C}}{C}=\overset{\overset{|}{C}}{C}-C-C$; ; $C-C=\overset{\overset{|}{C}}{C}-C$

MINILAB 1

Chemicals and supplies per 10 students:

0.2M bromine/carbon tet.	5 ml (0.16 g Br_2, 8 g CCl_4)
1,2-dibromoethane	55 g (25 ml)
1-pentanol (n-amyl alcohol)	20 g (25 ml)
1% potassium permanganate	5 ml (0.05 g $KMnO_4$)
zinc, granular, 20-30 mesh	25 g
glass tubing, 5-6 mm, 4' lengths	3
wood splints	15

Comments: The gas can be collected using a pneumatic trough or a 1-ℓ beaker filled with water. Ethylene is flammable and should decolorize the permanganate and bromine solutions (see Part III, tests C-7 and C-19 for reactions). If the delivery tubes are to be constructed by the students, burners, flame spreaders and triangular files should also be provided.

EXPERIMENT 9

Time: 1-1½ periods

Prelab calculations: 16.8 g (21.8 ml) cyclohexane; 27.7 g (31.9 ml) toulene

Chemicals and supplies per 10 students:

cyclohexane	220 g (280 ml)
toulene	360 g (415 ml)
glass beads, 3-4 mm diameter	300 g
or	
sponges, stainless steel	3-4 (cut into thirds or quarters)

Special apparatus: gas chromatograph(s)

Comments: If time is limited, the second distillation can be omitted. If the GC is to be run at a later time, students must be cautioned to keep their samples in tightly sealed containers to prevent evaporation. If there are not enough gas chromatographs to run all five samples, analysis of fractions 2 and 3 (and the graph in Calculation #1) can be omitted. We have used a 3% SE-30 Aeropak #30 column at 70° for the GC analysis, but almost any general purpose column (e.g. Apiezon L on Chromosorb) should suffice.

Students may require help with the HETP calculations. The Fenske equation can be written in logarithmic form as $n = \log(Z_A X_B / Z_B X_A)/\log\alpha$. Example: if a student measures 90% (by weight) cyclohexane in the initial distillate, the value of n is $\log\ [(90.8 \times 60)/(9.2 \times 40)]/\log 2.43 = 3.04$, since 90% cyclohexane is 90.8 mole percent and the initial mixture is 40 mole percent cyclohexane. Then the number of theoretical plates for the column is 3.04 - 1 = 2.04, and the HETP for a 15 cm column would be about 7.4 cm. Typical student results range between 80-90% cyclohexane. Reported HETP values for glass-bead packing are about 8-9 cm.

Topics for Report:

1. At a given distillation temperature, the distillate is richer than the distilland in the more volatile component. For example, the distillate collected in the 85-97° fraction is considerably richer in cyclohexane than was the liquid in the pot while it was being collected. If this distillate is redistilled, it will begin distilling at a temperature lower than 85°, and should yield a considerable amount of the 81-85° fraction.

2. "Stepping off" the graph gives approximate values of: (a) 40 mol% cyclohexane and (b) 62, 78, 88 and 93 mol% cyclohexane, respectively.

13

3. $Z_A/Z_B = \alpha^n(X_A/X_B) = 2.43^{(6 + 1)}(40/60) = 333.5$. Mole

percent cyclohexane = 100 x 333.5/334.5 = 99.7 mol% (assuming constant α).

4. Washing with fuming sulfuric acid, which removes the toluene as a soluble sulfonate derivative.

EXPERIMENT 10

Time: 1 period or less

Prelab calculations: 2.96 g trans-cinnamic acid; 10 ml Br_2/CCl_4

Chemicals per 10 students:

bromine	40 g (13 ml)
carbon tetrachloride	600 g (375 ml)
trans-cinnamic acid (Aldrich C8,085-7)	40 g
methylene chloride	400 g (300 ml)

Solution preparation: 2.0M Br_2/CCl_4 - 40 g bromine + carbon tet. to make 125 ml solution.

Comments: The addition funnel should be kept tightly stoppered throughout the addition unless the reaction is run under a hood; the solution is dense enough that pressure release is unnecessary. 5-10 ml of methylene chloride are needed for each washing, and the product should be pure white. Yields are high, averaging about 80%; the product melts around 201-2°, as expected for erythro-2,3-dibromo-3-phenylpropanoic acid. Most students can complete the experiment in 2 hours or less; it can be lengthened by requiring students to recrystallize a gram or so of their product from chloroform for the melting point determination, or by assigning Minilab 2.

Topics for Report:

1.

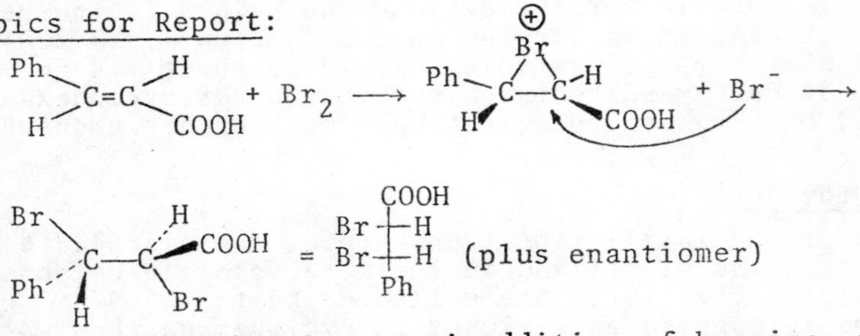

The reaction involves anti addition of bromine, yielding the erythro dibromide.

2. No. Add bromine to cis-cinnamic acid and see if it forms the threo-dibromide.

14

3.   cinnamic acid, bromine, carbon tetrachloride

                              |reflux

   dibromide, carbon tet, (bromine), cinnamic acid)

                              |cool, filter, wash

          filtrate   ppt.
      ┌───────────────────────────┴──────────────
carbon tet, bromine,   dibromide
   cinnamic acid

4. (a)

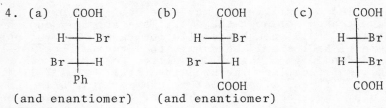

(and enantiomer)   (and enantiomer)

MINILAB 2

Chemicals per 10 students:

bromine	5.6 g (1.9 ml)
carbon tetrachloride	350 g (220 ml)
butter	1½ g
gasoline	2 ml
linseed oil	2 ml
mineral oil	2 ml
turpentine	2 ml
rubber cement	2 ml
vegetable oil	2 ml
vegetable oil, hydrogenated	1½ g

Solution preparation: 1.0M Br_2/CCl_4 - 5.6 g bromine in
carbon tet. to make 35 ml solution.

Comments: Turpentine, rubber cement, vegetable oil, linseed oil and oleomargarine should show considerable unsaturation; butter is slightly unsaturated; and the rest are
saturated or nearly so. Turpentine contains α- and β-
pinene; rubber cement contains about 7% polyisoprene or
similar unsaturated polymers; linseed oil, vegetable oil
and oleomargarine contain triglycerides of oleic, linoleic,
and other fatty acids; butter contains saturated and unsaturated fatty acid triglycerides. Students can work in
pairs; they should use 1.0M bromine/carbon tet rather than
the 0.2M solution indicated in test C-7; otherwise too
many drops are required.

EXPERIMENT 11

Time: 1½-2 periods

Prelab calculation: 16.8 g (19.4 ml) 2-methyl-3-butyn-2-ol

15

Chemicals per 10 students:

ethyl ether	450 g (630 ml)
magnesium sulfate, anhydrous	30 g
2-methyl-3-butyn-2-ol	220 g (250 ml)
(Aldrich 12,976-3)	
mercury(II) sulfate	13 g
potassium carbonate dihydrate	325 g
(or 1½-hydrate)	
sodium chloride	300 g
3M sulfuric acid	1 l (167 ml conc. H_2SO_4)

Special apparatus: infrared spectrophotometer(s)

Comments: The codistillation may take over an hour; students often do not heat the distilland strongly enough to maintain a rapid enough distillation rate. Separation: about 20 g of NaCl should be enough to saturate the distilland, and 2 g of magnesium sulfate to dry the ether solution. Purification: The product distills at 137-141°C. Analysis: If IR spectrophotometers are not available for student use, the spectra in the back of this manual can be photocopied and distributed to students for interpretation. E.V. 2: The NMR spectrum of the product can be found in the Aldrich Library of NMR Spectra, 2, 116C.

Topics for Report:

1. The following significant bands (approx. wavenumbers in cm^{-1}) are present in both reactant and product spectra: O-H (3300); C-H (2950); C-O (1180). The reactant shows a ≡C-H stretch (3250) that is partly obscured by the O-H band, a C≡C stretch (weak, 2100), and a ≡C-H bend (650). The product has a strong C=O stretching band at 1700. Disappearance of the alkyne bands coupled with appearance of the carbonyl band clearly shows the conversion that has taken place.

2.   3-methyl-3-butyn-2-ol, mercuric sulfate,
                    sulfuric acid, water
                              |reflux
      3-hydroxy-3-methyl-2-butanone, sulfuric acid
        mercuric sulfate, water, (unreacted alkyne)
                              |codistill
          residue |distillate
      mercuric sulfate,   ketone, water, (alkyne),
    sulfuric acid, water | (sulfuric acid)
                              |salt out, extract
          aqueous | organic
      sulfuric acid   ketone, ether,
        (as salt)   (alkyne), (water)
                                        |

(Continued on Next Page)

```
                                        |dry
                       residue          |
             water (as hydrate)         |
                                        |
                                        |distill
                       forerun          |    distillate
                       ├─────────────   └────────────
                       ether,           3-hydroxy-3-methyl-2-
                       alkyne                  butanone
```

$$3. \quad \underset{\text{C}}{\underset{|}{\overset{O}{\overset{\|}{\text{C}}}-\overset{OH}{\overset{|}{\text{C}}}-\text{C}}} \xrightarrow{H^+} \underset{\text{C}}{\underset{|}{\overset{HO}{\overset{|}{\text{C}}}-\overset{OH}{\overset{|}{\overset{\oplus}{\text{C}}}}-\text{C}}} \xrightarrow[\text{shift}]{\text{methyl}} \underset{\overset{\oplus}{\text{C}}}{\underset{|}{\overset{HO}{\overset{|}{\text{C}}}-\overset{OH}{\overset{|}{\text{C}}}-\text{C}}} \xrightarrow{-H^+} \underset{\text{C}}{\underset{|}{\overset{HO}{\overset{|}{\text{C}}}-\overset{O}{\overset{\|}{\text{C}}}-\text{C}}}$$

4. (a) CH_3CHO;   (b) ![cyclohexane with OH and COCH3]; (c) $CH_3COCH_2CH_3$;

   (d) $CH_3COCH_2CH_3$

EXPERIMENT 12

Time: 1-1½ periods

Prelab calculations: 8.22 g (10.1 ml) cyclohexene; 24.2 ml chloroform

Chemicals per 10 students:

chloroform	450 g (300 ml)
cyclohexene	100 g (125 ml)
magnesium sulfate, anh.	65 g
methylene chloride	1.3 kg (1 1)
10M sodium hydroxide	900 ml (360 g NaOH)
tetrabutylammonium bromide	7 g
(Aldrich 19,311-9)	

Special apparatus: magnetic stirrers and stir bars

Comments: Efficient mixing is essential for good yields in this preparation. If desired, the experiment can be performed by shaking the reactants (in a 250 ml Erlenmeyer flask) in a 60° water bath under the hood for 45 minutes or more; this procedure is tedious, however. Separation: The positions of the layers are sometimes inverted, with chloroform going to the top (or even the middle!), so they should be tested to determine which layer is organic. About 75 ml (total) of methylene chloride should be used to extract the aqueous layers (emulsions should clear up on standing) and 4-5 g of drying agent is needed to dry the combined organic layers. Purification: After discarding a forerun that boils below 120°, the product is collected at 190-200°. E.V. 1: GLC has been carried out using silicone oil/Chromosorb W at 110°. The IR spectrum·is

17

reproduced in A. Ault, Techniques and Experiments for Organic Chemistry, 3rd ed. (Boston: Allyn & Bacon, 1979), p. 274; and in J. Am. Chem. Soc. 76, 6163 (1954).

Topics for Report:

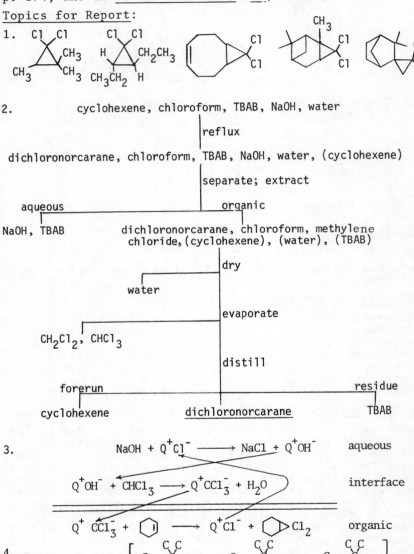

1.

2.   cyclohexene, chloroform, TBAB, NaOH, water

reflux

dichloronorcarane, chloroform, TBAB, NaOH, water, (cyclohexene)

separate; extract

aqueous   organic

NaOH, TBAB   dichloronorcarane, chloroform, methylene
                  chloride, (cyclohexene), (water), (TBAB)

dry

water

evaporate

CH_2Cl_2, $CHCl_3$

distill

forerun   residue

cyclohexene   dichloronorcarane   TBAB

3. $NaOH + Q^+Cl^- \longrightarrow NaCl + Q^+OH^-$ aqueous

$Q^+OH^- + CHCl_3 \longrightarrow Q^+CCl_3^- + H_2O$ interface

$Q^+ CCl_3^- + \bigcirc \longrightarrow Q^+Cl^- + \bigcirc Cl_2$ organic

4.

$$\xrightarrow[-H^+]{H_2O} \quad \text{CC=CCCC=C} \; + \; \text{HOCC=CCC=C}$$

(with structural fragments: CHO C / CC=CCCC=C and C C / HOCC=CCC=C)

EXPERIMENT 13

Time: 1 period

Prelab calculations: 33.0 ml acetic anhydride; 2.6 ml 48% fluoboric acid; 3.9 g triphenylmethanol; 1.57 g (1.8 ml) cycloheptatriene.

Chemicals and supplies per 10 students:

acetic anhydride	450 g (415 ml)
cycloheptatriene (Aldrich C9,920-5)	20 g (23 ml)
dimethyl sulfoxide or DMSO-d_6, etc.	7 g (6½ ml)
ethyl ether, anhydrous	650 g (910 ml)
48% fluoboric acid (Baker 9528)	50 g (33 ml)
methanol	120 g (150 ml)
potassium iodide	75 g
tetramethylsilane	0.15 g (0.2 ml)
triphenylmethanol (Aldrich 13,484-8)	50 g
NMR tubes	10

Special apparatus: NMR spectrometer(s) (optional)

Comments: The preparative part of the experiment can be finished in about 2 hours. If NMR spectrometers are not available for student use, that part of the experiment can dry-labbed by distributing copies of the NMR spectrum in the back of this manual. Students should allow time for the precipitate of tropylium fluoborate to consolidate before filtering, and must take care to keep water away from the product. Two 10-15 ml portions of ether should be sufficient for washing this product. If protic DMSO is used as the NMR solvent, the spectrum can be scanned only at low field (about δ5-10) to avoid the solvent signal at δ2.6. A signal at δ7.1 is probably due to hydrolysis of the product to form tropyl alcohol. Acetonitrile (δ2.0) should also be a suitable NMR solvent. Part B: Students can prepare their own KI solutions by dissolving 5 g of KI in about 3½ ml of water. 10 ml of hot water is sufficient to dissolve the tropylium fluoborate; the water can be heated on a hot plate away from the work area so students will not be tempted to use a burner for that purpose while ether is in use. Tropylium iodide forms beautiful red crystals; it may be quite impure due to reaction with the water (see Topic 2). E.V. 1: Use about 10^{-4}M tropylium in 0.1M HCl and 10^{-3}M benzene in ethanol. Tropylium U.V. maxima are at longer wavelengths than those of benzene (about 217 and 274 nm compared to 184 and 204 nm) because of the longer conjugated system. E.V. 2: The IR spectrum of tropylium ion has only four peaks of appreciable intensity, at 3020, 1480, 680 and 650 cm^{-1} (J. Am. Chem. Soc. 76, 3203 (1954)).

Topics for Report:

1. (a) The NMR spectrum of tropylium fluoborate indicates that all the ring protons are equivalent, which is not true of cycloheptatriene; also the large downfield shift indicates a high ring current, characteristic of aromatic compounds.
   (b) $\delta 2.25$ - 7,7-protons; $\delta 5.4$ - 1,6; $\delta 6.15$ - 2,5; $\delta 6.55$ - 3,4.

2. Triphenylmethanol hydrolyzes to trityl alcohol according to the reaction $Ph_3C^+BF_4^- + H_2O \rightarrow Ph_3COH + HBF_4$. The impurities in tropylium iodide are tropyl alcohol (C_7H_7OH) and probably some ditropyl ether $(C_7H_7)_2O$.

3. All are aromatic except 1 (4 pi electrons), 4 (5 pi electrons in ring), 5 (only 12 pi electrons in correct orientation for resonance; the other 6 are in the ring plane); 8 (4 pi electrons); and 9 (12 pi electrons).

4. Interference between the hydrogen atoms illustrated makes a planar structure difficult to attain without distorting bond angles. Deviation from planarity reduces the resonance energy.

EXPERIMENT 14

Time: 1 period

Prelab calculations: 1.22 g triphenylmethane, 5 ml Br_2/CCl_4

Chemicals and supplies per 10 students:

bromine	10.4 g (3½ ml)
carbon tetrachloride	300 g (190 ml)
cyclohexene	2 g (3 ml)
petroleum ether (60-75°) or hexane	100 g (150 ml)
1M sodium hydroxide (for gas traps)	300 ml (12 g NaOH)
toluene	45 g (50 ml)
triphenylmethane (Aldrich 10,130-3)	15 g
zinc, granular, 30 mesh	8 g
glass beads, 6 mm	10
light bulbs, unfrosted, 100 W, with power cord	4-5
pipets, volumetric, 5 ml	10
rubber stoppers, #4, 1 hole	10
rubber stoppers, #4, solid	10
rubber tubing, 3/16" I.D.	40 cm (4 cm lgths)
test tubes, sidearm, 25 x 200 mm	10

Solution preparation: 1.0M Bromine/carbon tetrachloride - Mix 10.4 g bromine with enough carbon tet. to make 65 ml of solution.

Comments: Some students may need more than one period to complete this experiment, because of their unfamiliarity with semimicro techniques. An ordinary 25 x 200 mm test

tube provided with a 2 hole rubber stopper and a bent glass
tube (for the gas outlet) can be substituted for the
sidearm test tube. The bromine solution can be measured
into a 5-ml volumetric pipet using the pipetting apparatus
illustrated in Figure 2, OP-6; removing the bulb and adding
a rubber stopper converts it to the "addition pipet"
illustrated in the experiment. Ten ml of pet. ether should
be sufficient for recrystallization. If the product is
exposed to water (as by aspirator backup during solvent
evaporation) it will hydrolyze to triphenylmethanol, which
does not yield the free radical in the following step.
Part B: If the trityl bromide/toluene solution is
appreciably colored, it should be treated with decolorizing
carbon before zinc is added. Addition of zinc should
result in a deep orange-yellow solution of the free radical,
which partly decolorizes and yields a white precipitate of
trityl peroxide when air is admitted. Stoppering the tube
allows regeneration of the free radical by decomposition
of the dimer (with which it is in equilibrium), until it
has all reacted with atmospheric oxygen. E.V. 2: The
precipitate is trityl peroxide, m.p. 185°C.

Topics for Report:

1. $2Ph_3C\cdot + O_2 \rightarrow Ph_3COOCPh_3$ White precipitate forms
                         (white ppt)   upon shaking with air,
                                              orange color fades.

    $(Ph_3C)_2 \rightleftarrows 2Ph_3C\cdot$ Orange color is regenerated
       dimer   when air is excluded.
    (colorless)   (orange)

    The true structure of the dimer is Ph_3C 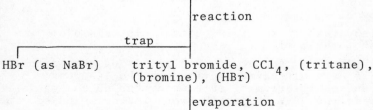 $=CPh_2$

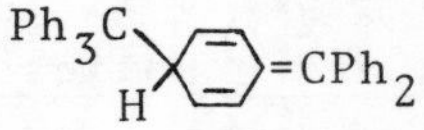

2. $Br_2 \xrightarrow{\text{light}} 2Br\cdot$ (chain-initiating step)

    $Ph_3CH + Br\cdot \rightarrow Ph_3C\cdot + HBr$
                                (chain-propagating steps)
    $Ph_3C\cdot + Br_2 \rightarrow Ph_3CBr + Br\cdot$

    $Br\cdot + Br\cdot \rightarrow Br_2$ (and other chain-terminating steps)

3.   tritane, bromine, CCl_4
                             │
                             │ reaction
                    ┌─────────┤
                    │ trap   │
        ┌───────────┘   │
    HBr (as NaBr)   trityl bromide, CCl_4, (tritane),
                     (bromine), (HBr)
                             │
                             │ evaporation
          (Continued on next page)

21

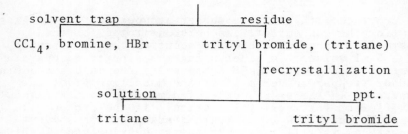

```
      solvent trap                    residue
  CCl₄, bromine, HBr          trityl bromide, (tritane)

                                      recrystallization

          solution                              ppt.

           tritane                         trityl bromide
```

4. 3 > 6 > 2 > 5 > 1 > 4

MINILAB 3

Chemicals per 10 students:

bromine (in 33 ml solution with CCl_4)	5.3 g (1.8 ml)
t-butylbenzene	11 g (13 ml)
carbon tetrachloride	500 g (315 ml)
ethylbenzene	11 g (13 ml)
isopropylbenzene (cumene)	11 g (13 ml)
toluene	11 g (13 ml)

Comments: The isopropylbenzene solution is decolorized almost immediately, ethylbenzene requires about a minute, toluene about 2 minutes, and t-butylbenzene is not decolorized after several hours. The order of free radical stability is: 3^0 benzylic > 2^0 benzylic > 1^0 benzylic > 1^0. The expected products are: $PhCBr(CH_3)_2$, $PhCHBrCH_3$, $PhCH_2Br$, and $PhC(CH_3)_2CH_2Br$.

EXPERIMENT 15

Time: 1-1½ periods

Chemicals and supplies per 10 students:

acetone	275 g (300 ml)
alumina, chromatography grade	200 g
0.025% iodine in hexane	1 ml (0.25 mg I_2)
magnesium sulfate, anhydrous	10 g
petroleum ether, 60-75° (or hexane)	1.5 kg (2.3 1)
10% potassium carbonate, aq. soln.	350 ml (35 g K_2CO_3)
sand	30 g
sodium chloride, satd. aq. soln.	350 ml (110 g NaCl)
tomato paste	50 g
burets, 25 ml (or other suitable columns)	10
glass wool	

Special apparatus: recording ultraviolet-visible spectrophotometer(s).

Comments: The "lycopene" in some brands of tomato paste appears to have already isomerized to the lycopene/ neolycopene equilibrium mixture, and thus no change is observed in the UV-VIS spectrum when iodine is added. We

22

have used Hunt's tomato paste with satisfactory results,
however. About ½ gram of drying agent is needed to dry
the extract; care should be taken to prevent overheating
during evaporation or exposure of the extract to sunlight.
We have used 25 ml burets (with teflon stopcocks) having
an inner diameter of about 11 mm for the chromatography.
About 15 g of alumina is needed for each column; with
larger diameter columns the quantities of alumina and
eluents should be increased. Most students can finish
the extraction and chromatography in one period, but they
may have to come back later to run the spectrum. If
possible, the spectrum should be run within a few days
since the extract decomposes in time. Students who take
sufficient care with the extraction and chromatography
will observe a large increase in absorbance at 360 nm
combined with a less dramatic decrease in absorbance in
the visible region (see spectra in the back of this manual).
The undiluted eluent can ordinarily be used for the UV
region on a 0-1 or 0-2 absorbance scale, but 2-fold or
4-fold dilution may be necessary in the visible region.

Topics for Report:

1. Using Beer's law, $c = A/\varepsilon b = A/(1.86 \times 10^4)(1.00)$ for
a 1 cm cell. The concentration is usually on the order
of $10^{-4}M$, with the undiluted solution have an absorbance
of 2-3 at 473 nm.

2. Decrease in absorbance of the bands in the visible
region; increase in absorbance at 360 nm, which is the
so-called "cis-peak" present in the spectra of most
cis-carotenoids. The bands in the visible region migrate
to slightly shorter wavelengths due to inhibition of
resonance in the cis isomer.

3. Extended conjugation allows low-energy pi→pi*
transitions, which are excited by wavelengths in the
visible region of the spectrum.

4. $ClCH_2CO_2Et + NaOEt \rightarrow Cl\overset{\ominus}{C}HCO_2Et + EtOH$

$$RC \overset{O}{\underset{}{\parallel}} CH_3 + Cl\overset{\ominus}{C}HCO_2Et \rightarrow R\overset{O\ominus}{\underset{H_3C \ \ Cl}{C-CHCO_2Et}} \rightarrow$$

$$R\overset{O}{\underset{CH_3}{C-CHCO_2Et}} + Cl^- \qquad R = $$

The reaction illustrates the Darzens condensation.

5. Both 7-cis and 11-cis-lycopene should
exhibit methyl-hydrogen steric interactions that would tend to inhibit

23

conjugation and decrease stability; other isomers such as 13-cis lycopene do not have a methyl group in a position to interact.

EXPERIMENT 16

Time: 1-1½ periods

Prelab calculations: 14.82 g (18.3 ml) 1- and 2-butanol; 28.3 ml 48% HBr

Chemicals per 10 students:

1-butanol	95 g (120 ml)
2-butanol	95 g (120 ml)
calcium chloride, anhydrous	10 g
48% hydrobromic acid	550 g (370 ml)
5% sodium bicarbonate, aq. soln.	150 ml (7.5 g NaHCO$_3$)
~1M sodium hydroxide (for gas traps)	1 1 (40 g NaOH)
sulfuric acid, conc.	425 g (230 ml)

Comments: About 10-15 ml each of 5% NaHCO$_3$ and water should be used to wash the alkyl bromide, and half a gram of calcium chloride should be sufficient for drying in most cases. The optimum yields of alkyl bromide should be obtained using 12 ml of sulfuric acid (just over 1 mole per mole of alcohol) with 1-butanol and 6 ml with 2-butanol. Because of variable losses during the final distillation, crude yields should be used for comparison purposes.

Topics for Report:
1. 12 ml (about 1 equivalent) sulfuric acid for 1-butanol, 6 ml for 2-butanol. Using 1 equivalent of the acid insures that all of the alcohol is protonated and results in the maximum rate for the reaction conditions. However, any rate increase for 2-butanol is offset by the occurrence of side reactions, such as E1 dehydration to 2-butene. (Elimination is less likely for 1-butanol, which reacts primarily by direct bimolecular substitution without forming a carbonium ion.)

2. (a) $\text{CCCC-OH} \xrightarrow{\text{H}^+} \text{CCCC-} \overset{\oplus}{\text{OH}}_2 \xrightarrow{\text{Br}^-} \left| \begin{array}{c} \overset{\text{H}}{\underset{\text{CCC} \;\; \text{H}}{\overset{\delta -}{\text{Br}} \text{-}\text{-}\text{-}\underset{\bigwedge}{\text{C}}\text{-}\text{-}\text{-}\overset{\delta +}{\text{OH}}_2}} \end{array} \right| \rightarrow$

$$\text{CCCC-Br} + \text{H}_2\text{O}$$

$$\underset{\text{CCCC}}{\overset{\text{OH}}{|}} \xrightarrow{\text{H}^+} \underset{\text{CCCC}}{\overset{\oplus \text{OH}_2}{|}} \xrightarrow{\text{-H}_2\text{O}} \overset{\oplus}{\text{CCCC}} \xrightarrow{\text{Br}^-} \underset{\text{CCCC}}{\overset{\text{Br}}{|}}$$

(b) The reaction of 1-butanol, since bromide ion participates in the rate-determining step.

24

3. 1-butanol:

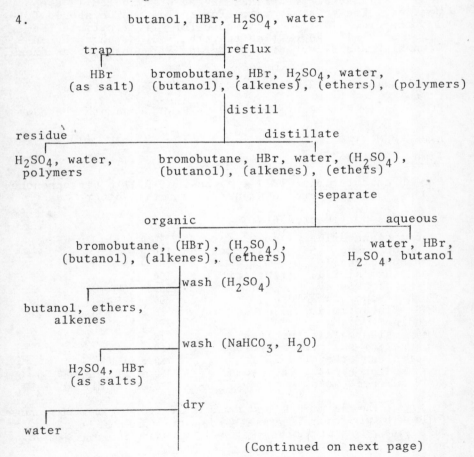

2-butanol:

(and other minor products resulting from carbonium
ion rearrangements and polymerization)

4.　　　　　　　butanol, HBr, H_2SO_4, water
　　　　　　　　　　　　|
　　trap　　　　　　　　| reflux
　　　|
　　HBr　　　　bromobutane, HBr, H_2SO_4, water,
(as salt)　　(butanol), (alkenes), (ethers), (polymers)
　　　　　　　　　　　　|
　　　　　　　　　　　　| distill
　　　　　　　　　　　　|
residue　　　　　　　　　　distillate
　|
H_2SO_4, water,　　bromobutane, HBr, water, (H_2SO_4),
　polymers　　　　　(butanol), (alkenes), (ethers)
　　　　　　　　　　　　　　　|
　　　　　　　　　　　　　　　| separate
　　　　　　　　organic　　　　　　　　　　　　aqueous
　　　　　　　|
bromobutane, (HBr), (H_2SO_4),　　　　　water, HBr,
(butanol), (alkenes), (ethers)　　　　H_2SO_4, butanol
　　　　　　　　　　　| wash (H_2SO_4)
　　　　|
butanol, ethers,
　alkenes
　　　　　　　　　　　| wash ($NaHCO_3$, H_2O)
　　　　　　|
　H_2SO_4, HBr
　(as salts)
　　　　　　　　　　　| dry
　|
water
　　　　　　　　　　　　　　(Continued on next page)

25

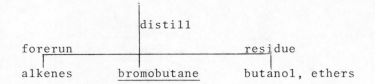

```
                        distill
    forerun                              residue
  ┌──────────────────────┴──────────────────┐
  alkenes         bromobutane        butanol, ethers
```

MINILAB 4

Chemicals per 10 students:

1-chlorobutane	5 g (6 ml)
2-chlorobutane	5 g (6 ml)
2-chloro-2-methylpropane (t-butyl chloride)	5 g (6 ml)
1% silver nitrate/ethanol (0.9 g $AgNO_3$ + 90 ml ethanol)	90 ml
15% sodium iodide/acetone (12 g sodium iodide + 85 ml acetone)	90 ml

Comments: The alkyl halides may be distributed in dropper bottles. The relative rates for sodium iodide should follow the order $1^0 > 2^0 > 3^0$ and those for silver nitrate the order $3^0 > 2^0 > 1^0$, suggesting that the iodide reacts by an S_N2 mechanism and silver nitrate by an S_N1 mechanism, as shown:

$$RC\ell + NaI \longrightarrow\!\mid \overset{\delta-}{I} \text{---} R \text{---} \overset{\delta-}{C\ell} \mid\!\longrightarrow RI + NaC\ell \text{ (precipitate)}$$

$$RC\ell + Ag^+ \rightarrow R^+ + \underline{AgC\ell}; \quad R^+ + NO_3^- \rightarrow R\text{-}NO_3 \text{ (and other products)}$$

EXPERIMENT 17

Time: 2 periods

Prelab calculations: 1.34 g magnesium; 9.11 g benzophenone; 7.85 g (5.25 ml) bromobenzene; 26 ml ethyl ether

Chemicals per 10 students:

acetic anhydride	140 g (130 ml)
benzophenone	120 g
bromobenzene	100 g (67 ml)
calcium chloride (or Drierite)	300 g
ethanol, 95%	300 g (380 ml)
ethyl ether, absolute	600 g (850 ml)
48% fluoboric acid, aq. soln.	30 g (20 ml)
iodine	1 g
magnesium metal, turnings	18 g
magnesium sulfate, anhydrous	30 g
petroleum ether (60-75°)	750 g (1.1 ℓ)
5% sodium bicarb., aq. soln.	350 ml (17.5 g $NaHCO_3$)
3M sulfuric acid	**270 ml** (45 ml conc. H_2SO_4)

Comments: The glassware is best dried in an oven, but an acetone rinse followed by air drying is generally adequate. Care must be taken during the reflux to prevent moisture from getting into the apparatus

26

through loose joints. Part B: A good stopping place is just after the addition of benzophenone, since the subsequent reflux can be omitted if the solution is allowed to stand overnight or longer. If the solution is not tightly stoppered much of the ether may evaporate; it should be replenished to facilitate the subsequent separation. About 25 ml of water and 25 ml of sodium bicarbonate are used in the washing, and the ether layer can be dried with 2 g of magnesium sulfate or another suitable drying agent (not calcium chloride). Petroleum ether is used to remove the by-product, biphenyl. Approximately 12 ml of 2:1 pet. ether/95% ethanol per gram of triphenylmethanol is needed in the recrystallization. Part C: The trityl fluoborate is orange in color and tends to darken on standing. E.V. 1: On adding water to the salt its color is discharged and triphenylmethanol precipitates. The U.V. spectrum has λ_{max} at 430 and 405 nm ($\log\epsilon$ = 4.6 for both). E.V. 4: The IR spectrum of triphenylmethanol is reproduced in the Aldrich Library of Infrared Spectra, 2nd ed., 604B.

Topics for Report:

1. $PhMgBr + H_2O \rightarrow PhH + Mg(OH)Br$ (from traces of moisture in the atmosphere)

    $PhMgBr + PhBr \rightarrow Ph\text{-}Ph + MgBr_2$

2.   magnesium, bromobenzene, ether
                                    | reflux

    PhMgBr, Mg, ether, (bromobenzene), (benzene), (biphenyl)
                                    | addition, reflux,
                                    | hydrolysis

    triphenylmethanol, ether, Mg^{2+}, Br^-, H_2SO_4, water,
        (bromobenzene), (benzophenone), (biphenyl), (benzene)
                                    | separate,
                                    | wash

    aqueous _____ organic

Mg^{2+}, Br^-, H_2SO_4 triphenylmethanol, ether, (bromo-
                                    benzene), (benzophenone), (biphenyl),
                                    (benzene), (water)
                                                        | dry
            water |
                        _____| evaporate
                ether, benzene

(Continued on next page)

                                                              wash, (peth),
                                                              filter
        biphenyl, benzene, PhBr
            biphenyl, benzophenone ──────────── recrystallize
                                    triphenylmethanol

3. PhMgBr + PhCPh → (Ph-C-Ph)MgBr $\xrightarrow{H_2O}$ Ph-C-Ph + Mg(OH)Br
        ‖O   O⁻ ⊕   OH
                        Ph   Ph

    PhMgBr + PhC-OEt → (PhC-Ph)MgBr → Ph-C-Ph + Mg(OEt)Br
         ‖O   O⁻ ⊕   ‖O
                         OEt

    (followed by reaction with more PhMgBr as in the first
     mechanism)

4.
(a)
            Ac₂O, AlCl₃
PhH ──┬──────────────────→ PhCOCH₃ ┐ $\xrightarrow[H^+]{H_2O}$ Ph-C-CH₃
      │ Br₂, FeBr₃   │   OH
      └──────────→PhBr →PhMgBr ─────┘   Ph
                      Mg
                    ether

(b)
            Br₂, hν
PhCH₃ ─┬─────────→PhCH₂Br →PhCH₂MgBr ┐   OH
       │   Mg   │ $\xrightarrow[H^+]{H_2O}$ Ph-CH-CH₂Ph
       │ CrO₃, Ac₂O   ether   │
       └────────────────→PhCHO────────┘

(c)
        Br₂, hν
PhCH₃ ─────────→PhCH₂Br ┐ AlCl₃   Br₂, hν
                        ├────→PhCH₂Ph ─────────→PhCHBrPh
                 PhH ───┘

    $\xrightarrow{Mg}{ether}$ PhCH-MgBr $\xrightarrow{HCHO}$ $\xrightarrow[H^+]{H_2O}$ PhCH-CH₂
                   Ph   OH
                                                                  Ph

(d)
        NaBH₄   OH   Mg   PhCOCH₃   H₂O
PhCOCH₃ ─────→PhCH-CH₃ $\xrightarrow{HBr}$ $\xrightarrow{ether}$PhCH-MgBr ─────→ $\xrightarrow{H^+}$
(from a)   CH₃

                        OH
            CH₃CH-C-CH₃
                Ph Ph

28

5.

1.   NMe$_2$ 2. $\oplus$NMe$_2$ 3.   NMe$_2$

Me$_2$N— —C$\oplus$ Me$_2$N— —C   Me$_2$N$\oplus$— —C

4.   NMe$_2$ 5.   NMe$_2$ 6.   NMe$_2$

Me$_2$N$\overset{\oplus}{}$— —C   Me$_2$N— $\overset{\oplus}{}$—C   Me$_2$N— —C$\underset{\oplus}{}$

and six more like 4-6 but with charges at o and p positions on the other two rings. (Alternate Kekule structures for the benzene rings are not considered here.)

EXPERIMENT 18

Time: 1½-2 periods

Prelab calculations: 13.62 g camphene; 28.6 ml acetic acid.

Chemicals per 10 students:

acetic acid	390 g (375 ml)
acetone	160 g (200 ml)
camphene (Aldrich 19,640-1)	175 g (200 ml)
chromium trioxide	95 g
75% ethanol, aqueous	620 ml (490 ml 95%)
magnesium sulfate, anhydrous	15 g
potassium hydroxide, pellets	107 g
10% sodium carbonate, aq. soln.	200 ml (20 g Na$_2$CO$_3$)
sulfuric acid, conc.	150 g (81 ml)
6M sulfuric acid	20 ml (6.7 ml conc. H$_2$SO$_4$)

Solution preparations: Jones' reagent - dissolve 95 g CrO$_3$
in 81 ml conc. sulfuric acid and cautiously dilute with
water to 350 ml. 2.5M KOH/ethanol - dissolve 107 g of KOH
pellets (85% KOH) in 75% ethanol, make up to 650 ml with
that solvent.

Comments: Aldrich's 99% camphene works well in this experi-
ment--their 80% camphene is liquid at room temperature and
must be purified by distillation. Most students can finish
the preparative work in one period so that the camphor can
be dried in a dessicator before purification. The sublimation is tedious without special equipment, so students
can weigh the crude camphor and only purify a gram or so
for the melting point determination. A burner and gauze
can be used for the sublimation (with the apparatus in (OP-
24) Fig. 1) but an oil bath is much more satisfactory. The

29

melting point determined in an open tube is usually several degrees lower than that measured in a sealed tube.

Topics for Report:

1. $m(\text{impurities}) = (179 - T_m)/40$ mol/kg

Since there are $1000/152.2 = 6.54$ mols camphor in 1 kg camphor, $\text{mol\%}(\text{impurities}) = \dfrac{100\ m}{6.57 + m} \cong 100m/6.57$

2. isobornyl acetate: $(1\ g/196.3) \times 1000 = 5.09$ mmol
   KOH: 2.5 ml $\times 2.5$ M $= 6.25$ mmol (23% excess)

   isoborneol: $(1\ g/154.3) \times 1000 = 6.48$ mmol;
   $6.48 \times 2/3 = 4.32$ mmol CrO_3 needed

   CrO_3: 1.75 ml $\times (6.75\ g\ CrO_3/25\ ml$ Jones' reagent$) = 0.473\ g\ CrO_3$
      $(0.473\ g/100) \times 1000 = 4.73$ mmol (9.5% excess)

3.
(a)

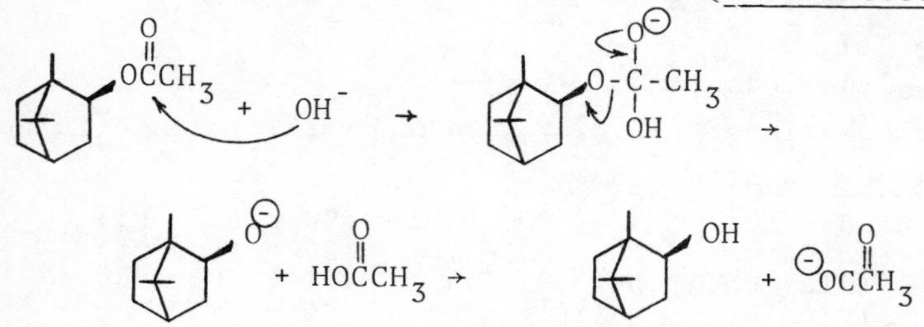

(b)

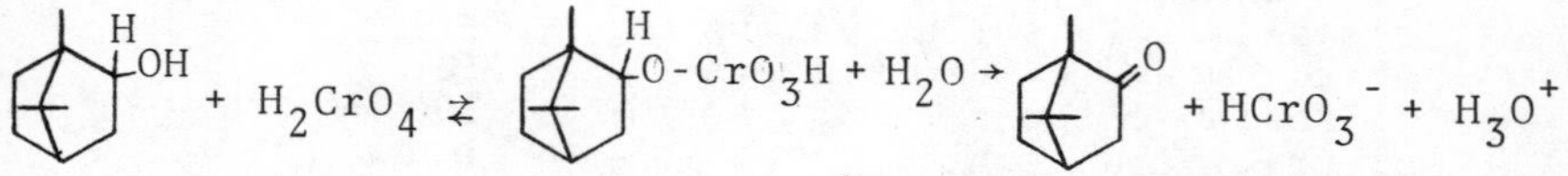

(Westheimer mechanism)

4.

(a)

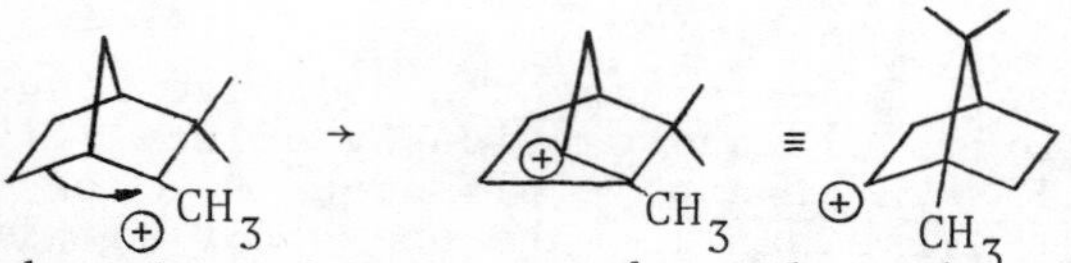

camphanyl cation   bornyl cation

(b)

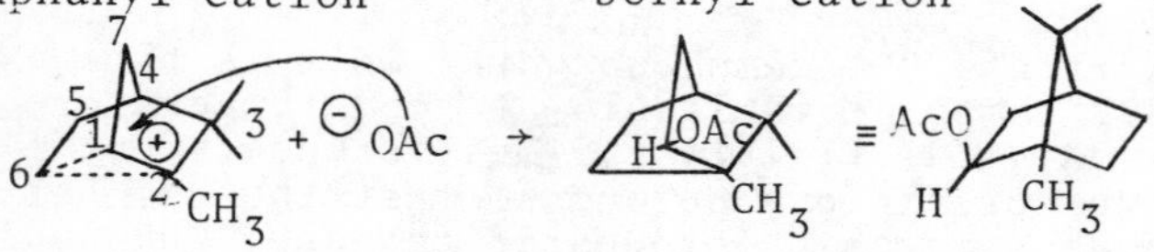

Attack by AcO^- on the other side of C-1 is prevented by the bridged C-6 carbon atom.

30

EXPERIMENT 19

Time: 1 period

Prelab calculations: 6.01 g (5.7 ml) acetic acid; 7.41 g (9.15 ml) 1-butanol

Chemicals per 10 students:

acetic acid	80 g (75 ml)
1-butanol	290 g (320 ml)
0.05% phenolphthalein (in 50% ethanol)	1 ml
0.40M sodium hydroxide	325 ml (5.2 g NaOH)
sulfuric acid, conc.	3 g (1.5 ml)

Comments: It is important to measure the volume of sulfuric acid accurately--otherwise the α values (especially with higher 1-butanol mole ratios) will be distorted. The acid can be dispensed in dropper bottles having calibrated droppers, with the number of drops per 0.1 ml indicated on the labels. Most students attain constant titration volumes after 45 minutes to 1 hour of refluxing. If Stark-Dean traps are not available, the apparatus illustrated in Figure 2 (OP-25c) can be used. The separator flask can be precalibrated by adding 7 ml of water and marking the level with a grease pencil. The most common error in using this apparatus is heating too rapidly, so that the drip tube becomes flooded. Students should attain close to 99% completion after water removal.

Literature values for K (Ind. Eng. Chem. 37, 968 (1945)) decrease measurably with an increase in the 1-butanol/acetic acid ratio (K = 2.87, 2.48, and 2.12 for mole ratios of 3, 5 and 10 respectively) but random variations in student results tend to obscure this effect.

Sample Calculations:

Example: Let V_0 = 15.0 ml, V_{equil} = 2.6 ml, V_{final} = 0.6 ml for 0.3 mol of butanol. Total volume $\cong$ 5.73 ml HOAc + 27.45 ml BuOH + 0.10 ml H_2SO_4 = 33.3 ml. Then V' = 9 x (2/33.3) $\cong$ 0.5 ml; V_e = 2.6 - 0.5 = 2.1 ml (corrected); α_e = (15.0 - 2.1)/15.0 = 0.86; K = 0.86^2/(0.14 x 2.14) = 2.47. Also V_f = 0.6 - 0.5 = 0.1 ml (corrected) so α_f = (15.0 - 0.1)/15.0 = 0.99.

(The number "9" in the V' equation comes from V_{acid} x (N_{acid}/N_{base}) = 0.1 x (36/0.4). A different factor should be used if the volume of acid or normality of the base are significantly different from 0.1 ml and 0.4N.)

Topics for Report:

1. (a) The percent yield (α x 100) should increase from about 62% for a 1/1 mole ratio to 92% for a 5/1 ratio, and

removing water should make the yield nearly quantitative. This illustrates Le Chatelier's principle in that adding more of a reactant and removing a product shifts the equilibrium to favor the products.

(b) See Comments above. The activity coefficients of the components may vary as their concentrations are varied; the true thermodynamic equilibrium constant is a function of activities rather than concentrations, so it may also vary.

2. Using a K value of 2.5:

$$K = \frac{0.75^2}{(1 - 0.75)(n - 0.75)} = 2.5 \qquad \underline{n = 1.65}$$

3. 1. Use RCOOH + diazomethane (CH_2N_2) or RCOCl + CH_3OH.

   2. Use PhCOCl + t-BuOH in the presence of a base (like dimethylaniline) to take up the HCl generated.

   3. Use acetyl chloride or acetic anhydride + m-cresol. The reaction with the acid chloride can be carried out in the presence of a base (pyridine or NaOH) to combine with the HCl generated. The reaction with the anhydride is usually catalyzed by a little sulfuric acid or by a base such as pyridine.

4. (a)

$$CH_3\overset{\overset{\displaystyle O}{\|}}{C}\text{-OBu} \underset{}{\overset{H^+}{\rightleftharpoons}} CH_3\overset{\overset{\displaystyle OH}{\|\oplus}}{C}\text{-OBu} \underset{}{\overset{H_2O}{\rightleftharpoons}} CH_3\underset{\underset{\displaystyle OBu}{|}}{\overset{\overset{\displaystyle OH}{|}}{C}}\text{-}\overset{\oplus}{O}H_2 \rightleftharpoons CH_3\underset{\underset{\displaystyle \overset{\oplus}{O}HBu}{|}}{\overset{\overset{\displaystyle OH}{|}}{C}}\text{-OH} \underset{}{\overset{-BuOH}{\rightleftharpoons}}$$

$$CH_3\overset{\overset{\displaystyle OH}{\|\oplus}}{C}\text{-OH} \underset{}{\overset{-H^+}{\rightleftharpoons}} CH_3\overset{\overset{\displaystyle O}{\|}}{C}\text{-OH}$$

(b) Use a large excess of water. Distilling off the alcohol would not work since the azeotropes contain more butyl acetate than 1-butanol. (The acetic acid could be removed by adding NaOH to convert it to sodium acetate, but then the reaction would not be acid catalyzed.)

(c) K(hydrolysis) = 1/K(esterification) ≈ 1/2.5 = 0.4

5. The titration would be erroneous otherwise. Returning the distillate to the reaction mixture brings it back to its original volume and yields the correct acetic acid concentration.

6.

$$HC\equiv CH \xrightarrow[Hg^{2+} \cdot H^+]{H_2O} CH_3CHO \xrightarrow[distill]{NaOH} CH_3CH=CHCHO \xrightarrow{H_2/Ni}$$

$$CH_3CH_2CH_2CH_2OH \overset{KMnO_4}{\longleftarrow}\longrightarrow CH_3COOH$$

$$CH_3COOH + CH_3CH_2CH_2CH_2OH \xrightarrow{H^+} CH_3COOCH_2CH_2CH_2CH_3$$

MINILAB 5

Chemicals per 10 students:

carboxylic acids (Table 2)	0.3 g each
ethanol, absolute	150 g (200 ml)
0.1M sodium hydroxide (standardized)	65 ml (0.26 g NaOH)

(These quantities assume that each student will run only one of the acids; otherwise, increase the quantities accordingly)

Special apparatus: pH meters

Comments: Students can work in groups of 3 or more, each being responsible for 1-2 acids, and pool their results. Measurements must be accurate for good results; burets are recommended for dispensing the liquids. The value of rho for this reaction in 75% ethanol is about 1.76 (from Chem. Rev. 53, 198 (1953)). The pK_a values estimated from the Hammett equation are (sigma values in parentheses): p-acetamidobenzoic: 6.35 (0.00); p-anisic: 6.8 (-0.27); benzoic: 6.35 (0.00); p-chlorobenzoic: 5.9 (0.23); p-hydroxybenzoic: 7.0 (-0.37); p-nitrobenzoic: 5.0 (0.78); p-toluic: 6.65 (-0.17). Students should conclude that electron-donating substituents (OH, OCH_3, CH_3) decrease acid strength by de-stabilizing the anion, and electron-withdrawing substituents (Cl, NO_2) increase acid strength by stabilizing it. In this reaction the acetamido group has virtually no effect on acidity, presumably because the electron-donating tendency of its nitrogen atom is nullified by the carbonyl.

Sample calculation: If the pH of the benzoic acid solution is found to be 6.19, its pK is calculated as follows:

c_o = 0.15/(122 x 0.020) = 0.061M benzoic acid (acid + salt);

$(A^-) = (OH^-) \cong 0.025M$; (HA) = c_o - (A^-) = 0.061 - 0.025 =

0.036M; pK = pH + log (HA)/(A^-) = 6.19 + log(0.036/0.025)

= 6.35

EXPERIMENT 20

Time: **may require about 1½ 3-hour lab periods.**

Chemicals and supplies per 10 students (working in pairs):

acetone	8 g	(10 ml)
2-butanone	8 g	(10 ml)
*chloroform	100 g	(67 ml)
2,4-dinitrophenylhydrazine	8 g	
95% ethanol	625 g	(800 ml)(see Comments)
ethyl acetate	18 g	(20 ml)
2-hexanone	8 g	**(10 ml)**
*methanol	30 g	(35 ml)
2-pentanone	8 g	(10 ml)
petroleum ether (~ 40-60°)	13 g	(20 ml)
*silica gel-G	33 g	

```
4-oz jars                              6
microscope slides (or pre-
 'coated silica TLC plates)    10
```

*omit if pre-coated plates are used

Solution preparation: 2,4-Dinitrophenylhydrazine reagent -
Mix 8 g DNPH with 40 ml conc. sulfuric acid and stir in
(slowly) 60 ml water to dissolve the solid. Add 200 ml
95% ethanol to the hot solution, mix well, filter if necessary.

Comments: Students can work in pairs, with each being
assigned a different unknown. Each student prepares a derivative of his/her unknown and two of the four known ketones. If microscope-slide TLC plates are used, each plate
can be spotted with these three derivatives. Two students
can use the same developing jar if necessary. One 4-oz jar
containing the silica gel slurry is sufficient for 10 or
more students. If 20 x 20 TLC plates are used, they can
each be cut into four 10 x 10 cm sheets; each sheet can be
spotted with two unknowns, the four known ketones separately, and a mixture of the four knowns. We have measured
the following approximate R_f values on microscope-slide TLC
plates: acetone, 0.24; 2-butanone, 0.34; 2-pentanone, 0.40;
2-hexanone, 0.45. The derivatives do not necessarily have
to be purified for the TLC analysis, so to save time and
ethanol you can require that only the unknown derivative be
recrystallized (for its melting-point determination).

Topics for report:

1. R_f values increase with molecular weight, decrease with
polarity. The less polar high-molecular-weight ketones are
less strongly adsorbed on the polar stationary phase and
more readily dissolved in the non-polar mobile phase, so
they travel farther in the mobile phase.

2.
$$CH_3\overset{\overset{O}{\|}}{C}\text{-}R \xrightarrow{H^+} CH_3\overset{\overset{OH}{|}}{\underset{\oplus}{C}}\text{-}R \xrightarrow{ArNHNH_2} CH_3\overset{\overset{OH}{|}}{\underset{R}{\overset{\oplus}{C}}}\text{-}\overset{}{N}H_2NHAr \xrightarrow{-H_2O}$$

$$CH_3\overset{\oplus}{\underset{R}{\overset{|}{C}}}\text{-NHNHAr} \xrightarrow{-H^+} CH_3\overset{}{\underset{R}{\overset{|}{C}}}\text{=NNHAr}$$

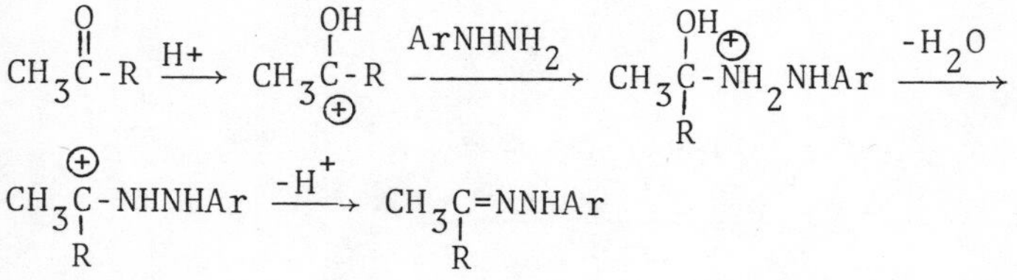

3. The acetone derivative, $(CH_3)_2C$=NNHAr, is more symmetri-
cal and its molecules pack together more regularly to form
a stable crystal lattice. (The more stable the lattice,
the more difficult it is to break apart and the higher the
melting point.)

4. (a) 2,4-dinitrophenylhydrazine: 0.20 g/198 = 1.0 mmol
(see D-3); (b) acetone: 0.25 g/58 = 4.3 mmol, 330%;
2-butanone: 3.5 mmol, 250%; 2-pentanone, 2.9 mmol, 190%,

34

2-hexanone, 2.5 mmol, 150%.

MINILAB 6

Chemicals and supplies per 10 students:

benzaldehyde	150 ml
sand	400 g
pipe cleaners	20

The product is benzoic acid (m.p. 122^0). Proof of identity should include a chemical test to establish that it is a carboxylic acid (**dissolves in 5% NaHCO₃**) and a mixture melting point, IR spectrum, or derivative preparation.

Reaction: $PhCHO + \frac{1}{2}O_2 \longrightarrow PhCOOH$

EXPERIMENT 21

Time: 1 period +

Prelab calculations: 4.09 g m-toluic acid; 2.6 ml thionyl chloride; 2.74 g diethylamine hydrochloride; 33 ml 3M sodium hydroxide.

Chemicals per 10 students:

diethylamine hydrochloride	35 g
ethyl ether	550 g (780 ml)
magnesium sulfate, anh.	40 g
1M sodium hydroxide (gas trap)	750 ml (30 g NaOH)
3M sodium hydroxide	450 ml (54 g NaOH)
sodium lauryl sulfate	1.5 g
thionyl chloride	55 g (35 ml)
m-toluic acid (Aldrich T3,660-9)	55 g

Special equipment:

gas chromatograph(s) (optional)
magnetic stirrers and stirbars

Comments: Most students can finish the experiment in one 3-4 hour period, but may have to come in later to run the GLC. The B part reaction can be carried out quite satisfactorily in a 250 ml Erlenmeyer flask, by (1) mixing together the NaOH, diethylamine-HCl, and sodium lauryl sulfate with cooling; (2) adding the acid chloride in small portions, stoppering and shaking the flask after each addition (rubber gloves, hood) and cooling the flask before the next addition; (3) **shaking the loosely stoppered** flask gently over a steam bath (hood) for 15 minutes.
Separation: The use of a surfactant generally results in emulsions during the extraction step. You can warn the students of this possibility in advance or let them discover on their own what is happening and how to deal with it. The emulsion can be broken by saturation of the aqueous layer with salt, swirling, and standing for 15 min or so. Some extra ether can be added, if necessary.
Purification: The semimicro vacuum distillation can be

carried out without a manometer, if necessary, since there should be only one major component distilling at an elevated temperature. The receiver requires no external cooling --any residual traces of ether will then evaporate, so it should not be necessary to change receivers. Analysis: The gas chromatography was carried out on a 8' x ¼" silicone oil/Chromosorb P column at a helium flow rate of 60 ml/min and a column temperature of 250^0. The purity of the product is generally good, with only a single significant peak in the chromatogram. This analysis can be omitted, if desired. 75-80% yields are easily attainable, but most students average around 50%.

If the purification is carried out by column chromatography (E.V. 1), about 50-100 ml of petroleum ether is generally required for elution of the product. Students testing their product by E.V. 3 should be cautioned to start with very small amounts of the solution, in case there is an allergic reaction to deet or to any impurities in the product.

Topics for Report:

1. (a) $HCl + NaOH \rightarrow NaCl + H_2O$; $SO_2 + 2NaOH \rightarrow Na_2SO_3 + H_2O$

   (b) $SOCl_2 + 4NaOH \rightarrow Na_2SO_3 + 2NaCl + 2H_2O$

   $ArSO_2Cl + 2NaOH \rightarrow ArSO_2O^-Na^+ + NaCl + H_2O$

2.
m-toluic acid
thionyl chloride

reflux

gas trap
HCl, SO_2 m-toluoyl chloride, thionyl chloride,
(as salts)   (m-toluic acid), (HCl), (SO_2)

add NaOH, amine,
surfactant;
heat

deet, sodium m-toluenesulfonate, NaOH, surfactant, water, NaCl, Na_2SO_3, (diethylamine)

extract

aqueous layer   organic layer

sodium m-toluenesulfonate, NaOH,   deet, ether,
surfactant, water, NaCl, Na_2SO_4 (water), (diethylamine)

dry, filter,

water   (Continued on next page)

36

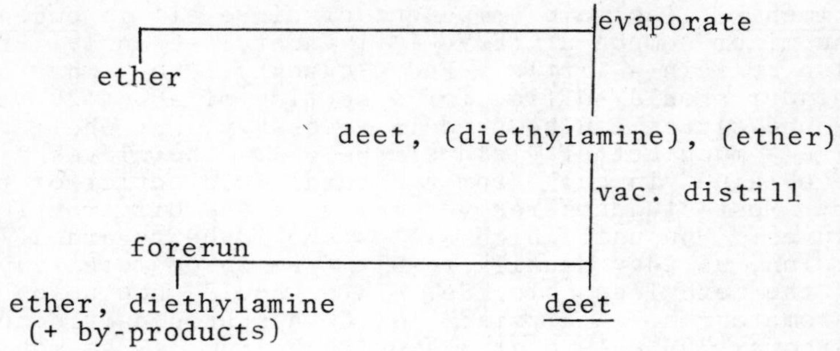

```
                                              evaporate
        ┌─────────────────────────────────────┐
ether   │                                     │
                        deet, (diethylamine), (ether)
                                     │
                                     │ vac. distill
            forerun                  │
        ┌────────────────────────────┘
    ether, diethylamine            deet
      (+ by-products)
```

3. 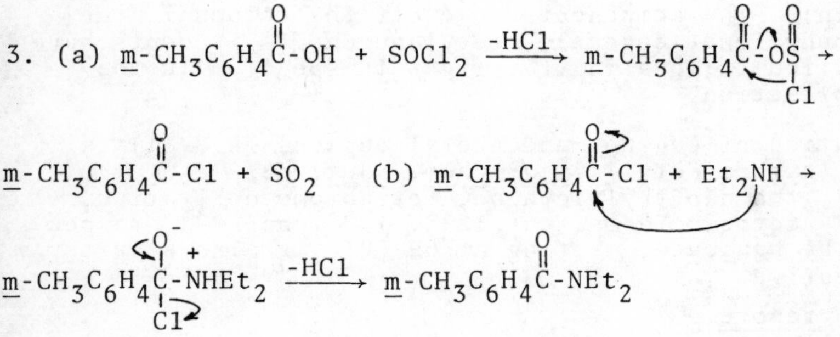 (a) $\underline{m}$-CH$_3$C$_6$H$_4$C(=O)-OH + SOCl$_2$ $\xrightarrow{-HCl}$ $\underline{m}$-CH$_3$C$_6$H$_4$C(=O)-O-S(=O)-Cl →

$\underline{m}$-CH$_3$C$_6$H$_4$C(=O)-Cl + SO$_2$ (b) $\underline{m}$-CH$_3$C$_6$H$_4$C(=O)-Cl + Et$_2$NH →

$\underline{m}$-CH$_3$C$_6$H$_4$C(O$^-$)(Cl)-$\overset{+}{N}$HEt$_2$ $\xrightarrow{-HCl}$ $\underline{m}$-CH$_3$C$_6$H$_4$C(=O)-NEt$_2$

4. PhNH$_2$ $\xrightarrow{n\text{-BuBr}}$ PhNHCH$_2$CH$_2$CH$_2$CH$_3$ $\xrightarrow{CH_3COCl}$ PhN(CH$_2$CH$_2$CH$_2$CH$_3$)-C(=O)CH$_3$

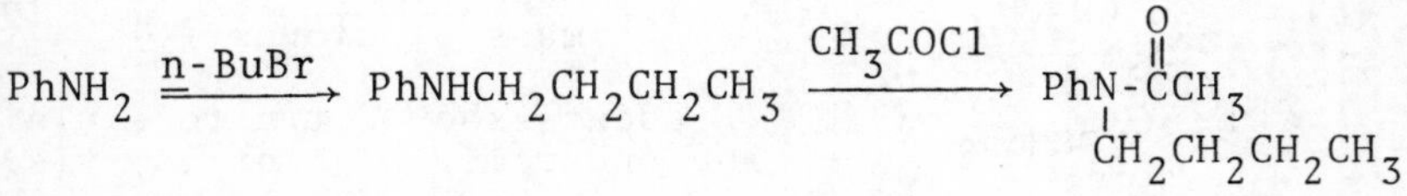

(Alternatively, the first step can be carried out by reductive amination of butanal; the second can utilize acetic anhydride rather than acetyl chloride.)

5. (a) 6 mmol excess thionyl chloride reacts with 24 mmol NaOH
     5 mmol excess m-toluoyl chloride r. w. 10 mmol NaOH
     25 mmol diethylamine-HCl r. w. 25 mmol NaOH
     25 mmol NaOH used up in acylation reaction
  Total: 84 mmol NaOH required

(b) 100 x (16/84) = 19% excess

EXPERIMENT 22

Time: 1 period +

Chemicals and supplies per 10 students:

cloves (whole)	125 g
magnesium sulfate	25 g
methylene chloride	860 g (650 ml)

Special equipment:

gas chromatograph(s)
infrared spectrophotometer(s)
Moulinex grinder(s)(or mortars and pestles)

<u>Comments</u>: The main component of clove oil is <u>eugenol</u> and
the minor component (E.V. 1) is acetyl eugenol. It may
also contain a little α-and β-carophyllene. One Moulinex
grinder should suffice for a section of about 20 students.
Ground cloves can be used if necessary, but whole cloves
give a much better yield of the oil. The cloves can often
be obtained in bulk from a natural food outlet or food co-
op. Most students recover 1-2 g of the oil from 10 g of
cloves. Unusually high yields should be regarded with sus-
picion, as they usually result from incomplete evaporation
of the methylene chloride. This should show up on the gas
chromatogram. <u>Analysis</u>: The GC has been carried out on a
Carbowax 20M column at 150° with helium as the mobile phase
(~40 ml/min). The components come off the column in the
order carophyllene, eugenol, acetyl eugenol.* Students have
little difficulty locating the right IR spectrum in the
Aldrich collection.

The separation of eugenol and acetyl eugenol (E.V. 1) is
described in more detail in J. <u>Chem</u>. <u>Educ</u>. **53, 263 (1976)**.
For E.V. 2, the naphthylurethane, tetrabromo derivative,
and aryloxyacetic acid melt at 122°, 118°, and 80°, respectively. The benzoate, melting at 69-70°, is also a suitable derivative.

<u>Topics for report</u>:

1. (a)

%oil = $\frac{\text{mass of oil (g)}}{\text{mass of cloves (g)}}$ x 100

(b) (Students should try to explain
any unusually high or low yield.)

eugenol

2. Eugenol shows IR bands at approximately the following
wave number values (cm^{-1}): 3500 (O-H), 1600 (C=C); 1270
(C-O, Ar); 1030 (C-O, R); 920 (=C-H bend); 820 and 790
(Ar-H bend). Acetyl eugenol can be detected by its C=O
band at 1760 cm^{-1}.

3. A typical commercial synthesis is as follows:

isoeugenol   vanillin

An alkaline solution of nitrobenzene can also be used to oxi-
dize isoeugenol. With stronger oxidants, the OH group must
be protected.

* A minor component may be observed between the eugenol and acetyl eugenol
  peaks.

4.

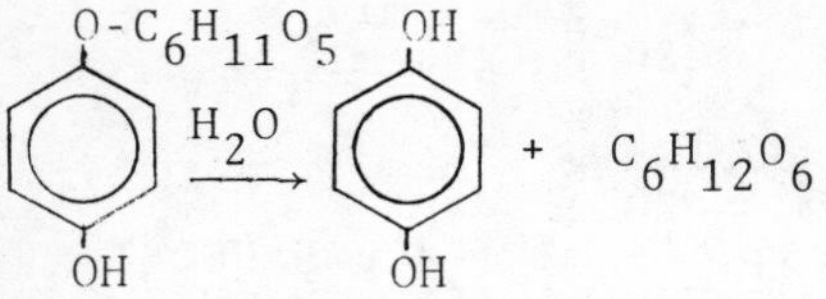

arbutin   compound X   glucose
         (hydroquinone)

Of the dihydroxybenzenes, only hydroquinone has all ring protons equivalent and would thus give an NMR spectrum with two sharp singlets.

MINILAB 7

Chemicals per 10 students:

hydrochloric acid (~3M)	5 ml
phenol	5 g
phthalic anhydride	3 g
sulfuric acid	4 g (2 ml)
sodium hydroxide (~3M)	40 ml (4.8 g NaOH)

Comments: One or two large $160°$ oil baths can be set up under a hood and used for an entire lab section. If burners are used for heating, the test tubes must be small enough so that a thermometer bulb can be immersed completely in the hot melt.

Reactions:

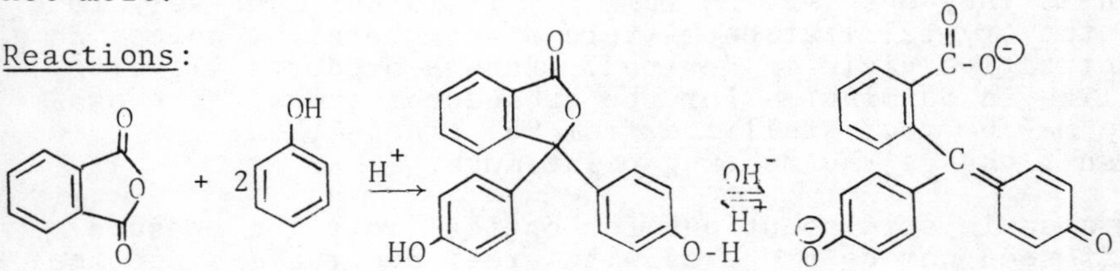

phenolphthalein
acidic form   basic form
(colorless)   (pink)

Adding NaOH converts the acidic form of phenolphthalein to its basic form (1 resonance structure illustrated), which absorbs visible (blue) light because it has a more extensive conjugated system than the acidic form.

EXPERIMENT 23

Time: 1 period

Chemicals per 10 students:

1% fructose solution (aq.)	7 ml (0.07 g fructose)
galactose	85 g
6M hydrochloric acid)	410 ml (205 ml conc. HCl)

(includes 250 ml for preparation of the Seliwanoff reagent)

methanol	20 g (25 ml)
nitric acid (conc.)	280 g (200 ml)
phenylhydrazine hydrochloride	13 g
resorcinol	0.125 g
sodium acetate trihydrate	20 g
sodium bisulfite, sat'd. aq.	15 ml (~4 g NaHSO$_3$)
2M sodium hydroxide	700 ml (56 g NaOH)

Solution preparation: Seliwanoff's reagent - dissolve 0.125 g of resorcinol in 250 ml of 6M hydrochloric acid.

Comments: EC-25 is galactose - this is the only hexose whose aldaric acid (mucic acid; galactaric acid) can be isolated by the procedure described - the others are too water soluble. The osazone melting point is about 202° (if performed correctly), which is consistent with structural units B and D. Of the four D-aldohexoses having these structural units, only D-galactose yields an optically in-active aldaric acid. (D-ketohexoses are eliminated by a negative Seliwanoff test.)

Students should prepare the osazone during the evaporation step of part A, which takes about 45 minutes over a boiling water bath. A 105 x 55 mm Pyrex evaporating dish is ideal for this but a 250-ml beaker will do. The reaction may be-gin suddenly, so students must watch the mixture carefully and remove the heat source (a burner is satisfactory) as soon as they observe the fuming. It is not necessary to let the crystallization mixture stand overnight unless a quantitative yield is desired. Enough product will crys-tallize in 30 minutes for the subsequent steps. The osa-zone may be recrystallized from 95% ethanol, but this doesn't change the melting point much.

The aldaric acid solutions for optical rotation measure-ments need not be prepared with great accuracy, since the rotation should be zero. The galactose solution should be prepared carefully, however (if the initial reading in a 1-dm tube is not close to +3°, a few drops of dilute HCl can be added and the rotation measured again when it sta-bilizes). It may be desirable to have students report their conclusion re the structure of EC-25 at the end of the laboratory period, then send them to the library to find the name that matches the structure and to compare their specific rotation values to the literature value (+80.2° for the equilibrium mixture of anomers).

Topics for report:

1. (a)

(b)

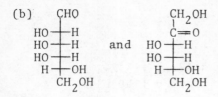

2. The optical rotation should be about $+3.2°$ in a 1 dm tube using a 1 g/25 ml galactose solution, giving a specific rotation:

$$\alpha \cong \frac{+3.2}{0.040 \cdot 1.0} = 80°$$

3. (a)

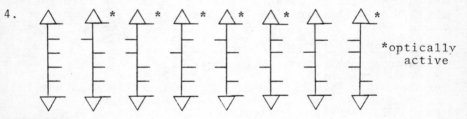

D-galactose   D-glucose

Although both glucose and galactose have the requisite gauche conformation between C-3 and C-4 hydroxy groups, the C-4 hydrogen in galactose can hydrogen-bond to the ring oxygen atom (as revealed by models), reducing its interaction with the receptor site (See J. Chem. Educ. 49, 171 (1972)).

(b) Heating provides enough energy to break the intra-molecular hydrogen bonds.

4.

*optically active

MINILAB 8

Chemicals per 10 students:

acetic acid	265 g (250 ml)
acetic anhydride	80 g (75 ml)
chloroform	22 g (15 ml)
cotton	7 g
sulfuric acid	4 g (2 ml)

Special equipment:

infrared spectrophotometer(s) (optional)

Comments: Students can also be requested to give a balanced equation for the reaction and to interpret the IR spectrum.

EXPERIMENT 24

Time: 2 periods

Chemicals per 10 students:

acetic acid	55 g (52 ml)
acetone	80 g (100 ml)
amino acid knowns	5 mg ea (in 1 ml water)
1-butanol	390 g (480 ml)
0.1% ammonia	100 ml (0.34 ml conc. NH_3)
cyclohexylamine	1 g (1 ml)
2,4-dinitrofluorobenzene	
(Aldrich D19,680-0)	0.25 g
dinitrophenyl-amino acids	2 mg ea (in 1 ml acetone)
dipeptide unknowns	0.05 g (5 mg ea)
ethanol, absolute	20 g (25 ml)
ethyl ether	175 g (250 ml)
6M hydrochloric acid	15 ml (7.5 ml conc. HCl)
hydrochloric acid, conc.	10 ml
ninhydrin (Aldrich 15,117-3)	0.25 g
3.5% sodium bicarbonate	2 ml (0.07 g $NaHCO_3$)

Solution preparations: Dinitrofluorobenzene solution -
combine 0.25 g dinitrofluorobenzene with 4.75 ml abs. ethanol. DNP-amino acid developing solvent - shake 300 ml 1-
butanol with 100 ml 0.1% aq. ammonia for 30 minutes or so, remove the aqueous layer in a separatory funnel. Amino acid developing solvent - combine 180 ml 1-
butanol, 45 ml acetic acid, and 75 ml water and mix thoroughly. Cyclohexylamine spray solution - mix 1 ml cyclohexylamine with 19 ml abs. ethanol. Ninhydrin solution -
mix 0.25 g ninhydrin with 7 ml acetic acid and 93 ml acetone.

Supplies per 10 students:

chromatography paper, Whatman #1,	
46 x 57 cm sheets	3
ovens	1 (at least 2 per lab section)
Petri dish lids (or bottoms),	
150 mm	1
pipets, graduated, 1 ml	
(or syringes)	10
polyethylene film (food wrap)	
sprayer	1
tweezers	1

Comments: The amino acids, dipeptides, and DNP-dipeptides
are available from United States Biochemical Corp. (P.O.
Box 22400, Cleveland, OH 44122), ICN Nutritional Biochemi-
cals (26201 Miles Road, Cleveland, OH 44128) and other
supply houses. These companies furnish kits containing a
selection of representative amino acids, peptides, etc.
We have used glycine successfully as an N-terminal group
(DNP-glycine has an R_f of 0.36 in the ammoniacal butanol

42

developer) but partial decomposition of the DNP derivatives results in a large dinitrophenol spot in the chromatography. Aspartic acid as an N-terminal group seems to give some difficulty. The dipeptides should be chosen so that the DNP derivatives of their component amino acids have significantly different R_f values--e.g. Phe-Leu would not be a good choice. We have used relatively inexpensive, readily available dipeptides like Ala-Gly, Ala-Leu, Ala-Phe, Ala-Val, Leu-Gly, Phe-Gly, Val-Gly etc., but many other combinations should work equally well.

Small test tubes (~10 x 75 mm) can be used in place of the conical centrifuge tubes if necessary. Dry air is not a necessity in the evaporations, since water or HCl is added in the subsequent steps, but some sort of trap is advisable to remove contaminants in the air line. (In any case, students should be made aware of the fact that lab air is not necessarily clean and dry and should learn how to purify it.) One trap or drying train for every 5-10 students should be sufficient. It is a good idea to demonstrate the tube-sealing technique at the start of the lab period, since many experiments are ruined by a hydrolysis tube that breaks during the heating period. Glass-distilled constant-boiling HCl (about 5.7M) has been recommended for the hydrolysis reactions, but ordinary 12M HCl diluted with distilled water seems to work fine. A 90-100° oven can be used for both hydrolyses if the time for the dipeptide hydrolysis is extended somewhat. Using a higher temperature for the DNP-dipeptide hydrolysis may result in charring.

The chromatography can be carried out by the method described in OP-18, using 600-ml or 800-ml beakers for development. A 46 x 57 cm sheet of chromatography paper can be cut into ten 11½ x 23 cm rectangles. Each sheet will accomodate three spots for the unknown amino acid or DNP-amino acid (using 1, 2, and 3 applications) and up to eight solutions of known amino acids or DNP amino acids. R_f values alone are not very reliable for identifying the DNP-amino acids, because they tend to streak. A black light should be kept handy to visualize any DNP-amino acid spots that are not visible to the naked eye. Spraying should be done under the hood. We have experienced some difficulty with the cyclohexylamine-ninhydrin combination, so unless the technique is checked out beforehand it might be best to use the traditional method of spraying with ninhydrin alone, which gives purple spots for all the amino acids.

Topics for Report:

1. Justification should include R_f and color comparisons from the amino acid chromatogram, with a rationale for eliminating other possible amino acids from consideration; and the student's reason for choosing one of them as the N-terminal amino acid, based on the results of the DNP-amino acid chromatogram.

2. (a) The equations under "Reactions and Properties" should be rewritten showing the side chains in the student's dipeptide.

(b) At high pH the C-terminal carboxyl group is ionized, making the DNP-peptide water soluble but not ether soluble. At low pH this group is not ionized (nor are any amino groups, being deactivated by DNP residues) so the DNP-peptide is less polar and therefore soluble in ether.

3.

O_2N—(ring, NO_2)—F + H_2NCH-$\overset{O}{\overset{\|}{C}}$-$NHCHCO^{\ominus}$ (with R and R' side chains) → O_2N—(ring, NO_2, F)—$\overset{\oplus}{N}H_2CHCNHCHCO^{\ominus}$ (with R, R')

$\xrightarrow{-HF}$ O_2N—(ring, NO_2)—$NHCH$-$\overset{O}{\overset{\|}{C}}$-$NHCHCO^{\ominus}$ (with R, R')

Nucleophilic aromatic substitution, bimolecular (S_NAr).

In alkaline solution the dipeptide zwitterion is converted to the anion shown, in which the amino group is unprotonated and can act as a nucleophile.

4. (a) -NHCHCO-, (CH$_2$)$_4$NH$_2$ [lysyl]
$\xrightarrow{DNFB}$ $\xrightarrow{H_2O}{HCl}$ $H_3\overset{\oplus}{N}CHCOO^{\ominus}$, (CH$_2$)$_4$NH-DNP

lysyl

-NHCHCO-, CH$_2$(ring)OH [tyrosyl]
$\xrightarrow{DNFB}$ $\xrightarrow{H_2O}$ $H_3\overset{\oplus}{N}CHCOO^{\ominus}$, CH$_2$(ring)O-DNP

tyrosyl

Side-chain amino and hydroxyl groups can be dinitrophenylated.

(b) They retain an unsubstituted amino group that is protonated at low pH, so the DNP derivative remains ionic and ether-insoluble.

(e.g., low pH form for Lys is $H_3\overset{\oplus}{N}CHCOOH$)
(CH$_2$)$_4$NH-DNP

5.   DNFB, dipeptide, NaHCO$_3$
                    |reaction

        DNP-dipeptide, DNFB (xs),
          NaHCO$_3$, (dipeptide)
                    | (Continued on next page)

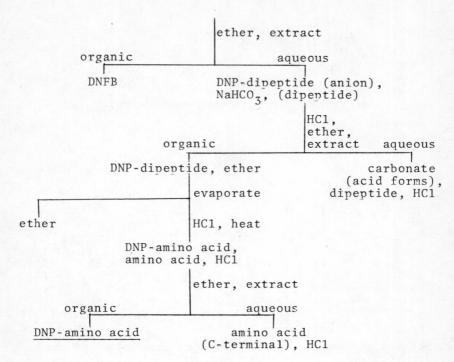

```
                              │ether, extract
          organic             │              aqueous
                              │
            DNFB              │     DNP-dipeptide (anion),
                              │     NaHCO₃, (dipeptide)
                              │
                              │
                              │HCl,
                              │ether,
            organic           │extract        aqueous
                              │
      DNP-dipeptide, ether    │              carbonate
                              │              (acid forms),
                              │evaporate     dipeptide, HCl
                              │
     ether                    │HCl, heat
                              │
            DNP-amino acid,   │
            amino acid, HCl   │
                              │
                              │ether, extract
          organic             │              aqueous
                              │
      DNP-amino acid          │     amino acid
                              │     (C-terminal), HCl
```

MINILAB 9

Chemicals per 10 students:

acetone	100 g (125 ml)
95% ethanol	120 g (150 ml)
3M hydrochloric acid	40 ml (10 ml conc. HCl)
skim milk	1.3 qt (1.2 l)
conc. nitric acid	3 g (2 ml)

About 3 g of casein can be obtained from 100 ml of cow's milk. The addition of nitric acid to casein should result in a bright yellow color. Students may also be asked to explain the chemistry of the xanthoproteic reaction (aromatic amino acid side chains are nitrated) and to report on the amino acid composition of casein or on the amino acid sequencing of bovine β-casein (Eur. J. Biochem. 25, 505 (1972))

PART II

ALTERNATE AND SUPPLEMENTARY EXPERIMENTS

EXPERIMENT 25

Time: 1 period

Chemicals and supplies per 10 students:

caraway oil	12 g (13 ml)
ethanol (95% or absolute)	200 g (250 ml)
methylene chloride	1.2 kg (900 ml)
petroleum ether (~60-80°)	1.7 kg (2.5 l)
silica gel, chromatography grade	400 g
spearmint oil	12 g (13 ml)
burets or chromatography columns	10

Special equipment: infrared spectrophotometer(s); polarimeter(s)

Comments: The essential oils can be obtained from Henry H. Ottens Mfg. Co., 1234 Hamilton St., Philadelphia, PA 19123 and from Fritzche D and O, 76 Ninth Ave., New York, NY 10011. Half of the class can be assigned one oil and half the other; they can be labeled "T'nim Oil" (caraway) and "Yawarac Oil" (spearmint) if you like. Caraway oil contains about 55% S-(+)-carvone, with the remainder being primarily R-(+)-limonene ($\{\alpha\}$ = +126°). Spearmint oil contains 50-60% R-(-)-carvone along with S-(-)-limonene and traces of α- and β-phellandrene. The odor difference between the two carvones is obvious to most observers, but about 10% are unable to tell the difference (see E.V. 4). The enantiomeric limonenes are also said to have intrinsically different odors, with the (+) form smelling like oranges and the (-) form like lemons (Science 172, 1044 (1971)).

Chromatography columns made from 22 mm (19 mm I.D.) glass tubing, drawn out at one end, are satisfactory. They should be packed with about 30 g of silica gel for separating of 2 g of the essential oil. If smaller diameter columns or burets are used, the quantities of adsorbent, essential oil, and eluents can be cut down in proportion to $(I.D.)^2$. The hydrocarbons (limonene and the phellandrenes)

should elute mainly in the second fraction and the carvones
in the fourth, with the third containing a mixture of these
components. The optical rotation can be measured using about ½ g of the carvone (or less, if necessary) in 10 ml of
ethanol, using a 1-dm polarimeter tube. To conserve ethanol, you may wish to run only one blank for every ten students or so, in which case the quantity of ethanol specified above can be reduced to about 150 ml (a little is
needed for rinsing). E.V.1: The physical properties are
recorded in the CRC Handbook and the spectra and gas chrom-
atograms are reproduced in J. Chem. Educ. 50, 74 (1973)
(labels on the IR and NMR spectra are interchanged). In
addition to the GLC conditions reported in that article, a
12' x 1/8" 30% GE SE-30/Chromasorb W column, or any relatively nonpolar column, can be used.

Topics for Report:

1. (a) Unless the chromatographic separation is very sharp
the carvones will not be optically pure, but their optical
rotations should have the correct signs and approximately
the right magnitude. The infrared spectra of the two carvones should be essentially identical, with C-H stretch
bands at 3080 to 2880 cm^{-1} , the C=O stretch at 1670 cm^{-1} ,
and =CH bending vibrations at 890 (s) and 800 (w) cm^{-1} .
The C=C stretch at 1640 cm^{-1} is partly obscured by the carbonyl band. Spectra can be found in the Sadtler and Aldrich collections as well as in the article cited above.
(b) Optical purity $\simeq 100(\alpha)/62°$. The calculated "optical
purity" may be greater than 100% if some limonene remains
in the carvones.

2.

R-(-)-carvone
spearmint oil

S-(+)-carvone
caraway oil

*chiral carbon atom

3. (a)

11 chiral carbons: 2^{11} = 1048
enantiomers

(b)

Carbon #	Configuration
3	R
5	S
7	R
8	R
9	S
10	S
12	S
13	R
14	S

(Configs. at C-17 and
C-20 not given)

4. (a)

CF_3CONH—$\overset{\displaystyle CH_3}{\underset{\displaystyle H}{|}}$—$COO$—$\overset{\displaystyle CH_3}{\underset{\displaystyle H}{|}}$—$CH_2CH_3$ CF_3CONH—$\overset{\displaystyle H}{\underset{\displaystyle CH_3}{|}}$—$COO$—$\overset{\displaystyle CH_3}{\underset{\displaystyle H}{|}}$—$CH_2CH_3$

    (S)   (S)   (R)   (S)

(b) Since there were two peaks of equal intensity in the gas chromatograph, the alanine must have been a racemic mixture of the R and S forms, yielding equal quantities of the diastereomers shown in (a). Biotic amino acids would be expected to consist primarily or entirely of one enantiomer, so the amino acid was probably not biotic.

MINILAB 10

Chemicals per 10 students:

D-(-)-tartaric acid	7 g
L-(+)-tartaric acid	7 g
DL-tartaric acid	7 g
meso-tartaric acid	7 g

Comments: The quantities specified assume that about 2 g of each acid is distributed to each group of four students. Since the D and meso forms are quite expensive, you may wish to assign them to only a few students, who can report their results to the rest of the class. The acids can be obtained from the Aldrich Chemical Co. with the DL and meso forms as hydrates. These may appear to melt at a lower temperature when the water of hydration is lost (e.g. the meso compound at $73-75^0$). If the melting-point tube is cut short so that the open end is below the heating block on a Mel-Temp, this water of hydration can be driven off and the true melting point obtained; alternatively the melting points can be obtained on a block and the top cover glass left off until the water evaporates. Since the specific rotations are quite low, about 2 g of the acid per 10 ml of aqueous solution is needed for accurate results. Infrared spectra can also be obtained if desired; they are repro- duced in the Aldrich Library of Infrared Spectra.

Interpretation:

D: COOH
   HO——H (S) m.p.: $171-4^0$
   H——OH (S) $\{\alpha\}_D^{20}$: -12^0
     COOH

L: COOH
   H——OH (R) m.p.: $172-4^0$
   HO——H (R) $\{\alpha\}_D^{20}$: $+12^0$
     COOH

meso: COOH
   H——OH (R) m.p.: 146-8
   H——OH (S) $\{\alpha\}$: 0
     COOH

DL: equal mixture   m.p.: 205
   of D and L $\{\alpha\}$: 0
   forms

48

Only the D and L forms are optically active, having equal but opposite rotations because they are enantiomers. The meso form is inactive because it has a symmetry plane and the DL racemate because it contains equal amounts of enantiomers whose rotations cancel out. The melting points of the D and L forms should be identical, but lower than that of the racemic mixture. Presumably the mirror-image molecules can pack together to form a more stable crystal lattice in the racemate. The meso compound has a different melting point because it is a diastereomer of the D and L forms.

EXPERIMENT 26

Time: 1-1½ periods

Prelab calculations: 3.43 g maleic anhydride, 6.80 g diene

Chemicals per 10 students:

ethyl ether (anhydrous)	90 g (125 ml)
3M hydrochloric acid	20 ml (5 ml conc. HCl)
maleic anhydride	45 g
methanol	550 g (700 ml)
petroleum ether (~60-80°)	85 g (125 ml)
α-phellandrene*	90 g (120 ml)
1M sodium hydroxide	50 ml (2.0 g NaOH)

Comments: The "unknown diene" is α-phellandrene--the other dienes named in the situation are either not readily available or did not give satisfactory results, in our experience. We have used Eastman Kodak α-phellandrene (P6406, b.p. 168-173°), but it is not listed in their latest catalog; Pfaltz & Bauer (375 Fairfield Ave., Stamford, CT 06902) list α-phellandrene, b.p. 175-6°. Most commercial α-phellandrene (about 60-70% pure) can be used without further purification in the quantity suggested in the Prelab calculations. According to David Todd (Experimental Organic Chemistry, Prentice-Hall, 1979, p. 257) the purity of commercial phellandrene can be determined by measuring its optical rotation in acetone (1g/10 ml); pure phellandrene has $\{\alpha\} = -177°$. Given this result, students can calculate the amount of the commercial chemical needed to provide about 35 mmol of α-phellandrene and use that amount rather than the suggested "50 mmol of diene". In any case, it is better to keep the diene in excess than the maleic anhydride. Methanol does not appear to solvolyze the anhydride appreciably if extended heating is avoided during the recrystallization. An ethyl ether-petroleum ether mixed solvent can also be used for recrystallization.

Student results in the hydrolysis are erratic, possibly due to partial decomposition, so this part can be omitted if time is limited. There may be some foaming during this reaction, so a 100-ml flask should be used for refluxing. If much of the solid remains undissolved during the

*Can also be obtained from MCB (Cat. #MX0256-10)

heating period, a little additional NaOH can be added--
using too much results in oil formation. If the acidified
reaction mixture yields a colloidal suspension, it may have
to stand overnight. The pure diacid melts at about 173°.
(Since this m.p. is not given in the experiment and is hard
to find in the literature, it can be used to evaluate student results.)

<u>Topics for Report</u>:

1.

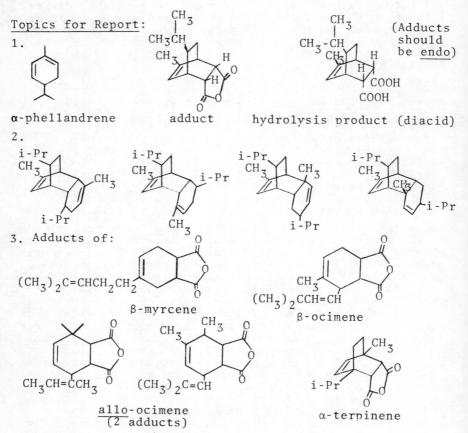

α-phellandrene   adduct   hydrolysis product (diacid)

2.

3. Adducts of:

$(CH_3)_2C=CHCH_2CH_2$

β-myrcene

$(CH_3)_2CCH=CH$

β-ocimene

$CH_3CH=CCH_3$

$(CH_3)_2C=CH$

allo-ocimene
(2 adducts)

α-terpinene

α-phellandrene, see Topic 1; β-phellandrene, no adduct possible, since double bonds cannot assume the necessary <u>cisoid</u>
orientation.

4.   maleic anhydride, phellandrene, ether

reflux

adduct, phellandrene (xs), ether,
        (maleic anhydride)

|(Continued on next page)

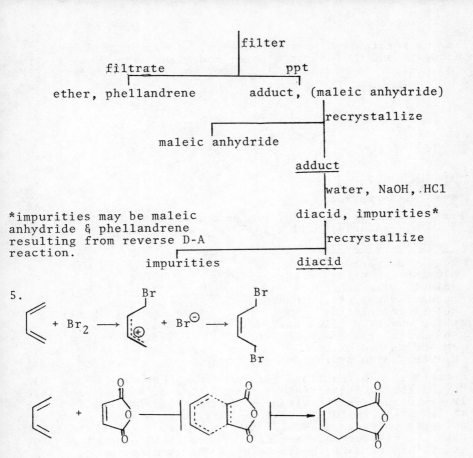

```
                              |filter
         filtrate            |            ppt
       |————————————————————|————————————————|
   ether, phellandrene            adduct, (maleic anhydride)
                                         |recrystallize
                       |—————————————————|
               maleic anhydride          |
                                       adduct
                                         |
                                         |water, NaOH, .HCl
  *impurities may be maleic          diacid, impurities*
  anhydride & phellandrene              |
  resulting from reverse D-A          |recrystallize
  reaction.              |—————————————|
                    impurities          diacid
```

5.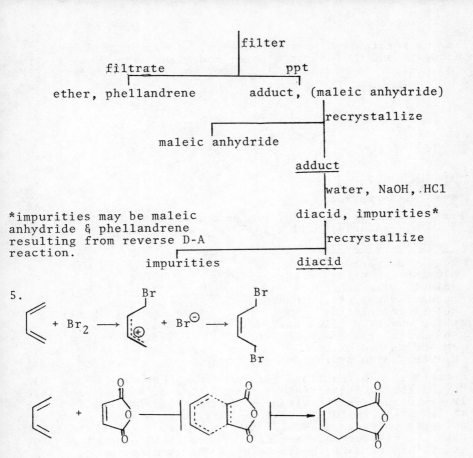

Both reactions involve attack at the ends (1,4-positions) of
a conjugated system by an electron-deficient species. The
bromine addition, however, involves bond heterolysis and
formation of a carbonium ion intermediate in a two-step
reaction. The Diels-Alder reaction involves bond homolysis
and the formation of a non-ionic activated complex in a one-
step, concerted reaction.

MINILAB 11

Chemicals per 10 students:

dioxane	68 g (65 ml)
ethyl ether (anhydrous)	35 g (50 ml)
furan	19 g (20 ml)
maleic anhydride	26 g

Comments: The reported melting point of the adduct is 125°.
It cannot be purified by recrystallization since it decomposes to the starting materials on heating. The adduct has

the unexpected <u>exo</u> stereochemistry, as established by R.B. Woodward and H. Baer in a paper you might wish to assign to your better students (<u>J. Am. Chem. Soc. 70</u>, 1161 (1948)).

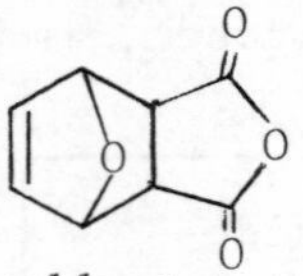

adduct

EXPERIMENT 27

<u>Time</u>: 1-1½ periods

<u>Prelab calculations</u>: 14.66 g aluminum chloride; 5.41 g (5.43 ml) anisole; 5.11 g (4.72 ml) acetic anhydride

<u>Chemicals per 10 students</u>:

acetic anhydride	65 g (60 ml)
aluminum chloride (Baker 0504)*	190 g
anisole (Aldrich 12,322-6)	70 g (70 ml)
magnesium sulfate, anhydrous	25 g
methylene chloride	660 g (500 ml)
petroleum ether (low boiling)	85 g (125 ml)
sodium chloride, sat. solution	350 ml (110 g NaCl)
3M sodium hydroxide	350 ml (42 g NaOH)
dilute sodium hydroxide (gas trap)	1 ℓ (≈50 g NaOH)

<u>Comments</u>: The reaction mixture turns a deep red-brown color during the addition; the aluminum chloride may not completely dissolve. Before cleaning the reaction flask with water, it is advisable to rinse it with a little ethanol or other suitable solvent--otherwise the water may react violently with the residue in the flask. About 25 ml each of 3M NaOH and saturated NaCl should be used for washing, and 1.5-2 g magnesium sulfate for drying. The product does not solidify during the distillation if no cooling bath is used for the receiver; a little methylene chloride can be used for transfers. The product can be purified by vacuum distillation, if desired.

<u>Topics for Report</u>:

1. $(CH_3CO)_2O + AlCl_3 \longrightarrow CH_3CO^+ \ AlCl_3(OAc)^-$

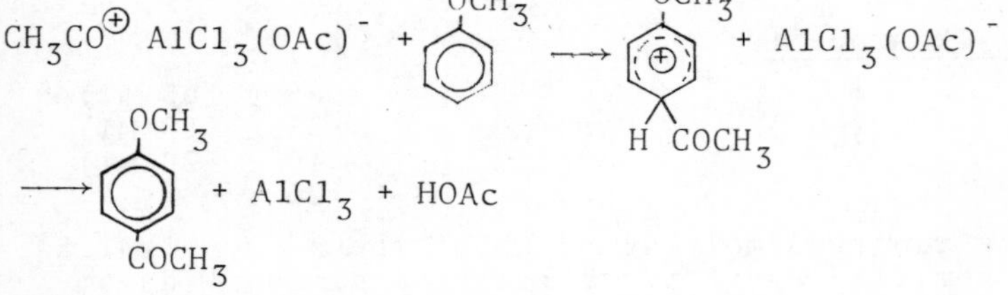

---

*The Baker Analyzed aluminum chloride is recommended; we had less satisfactory results with the Aldrich product.

2.   anisole, aluminum chloride,
           methylene chloride, acetic anhydride
                              |
                              | acetic anhydride;
                              | reflux
    trap |
    ┌─────────────────────────┘
    HCl (as NaCl)   p-methoxyacetophenone*, acetic acid*,
                         aluminum chloride, methylene chloride,
                         (anisole), (acetic anhydride), (HCl),
                         (by-products**)
                              |
                              | ice water
        aqueous | organic
    ┌─────────────────────────┴──────────────────┐
    AlCl₃ (as hydroxide),   p-methoxyacetophenone, methy-
                                 lene chloride, (anisole), (ace-
    acetic anhydride (as   tic acid), (by products)
    HOAc), acetic acid,
    HCl |
                                      | wash (NaOH, NaCl)
        aqueous | organic
    ┌─────────────────────────────────┴────────────┐
    acetic acid   p-methoxyacetophenone,
    (as NaOAc)   methylene chloride, water,
                                 (anisole), (by products)
    *as AlCl₃ complex | dry
                                   ┌────────┘
    **o-methoxyacetophenone,   water
    etc. | evaporate
                            ┌───────────────┘
                    methylene chloride
                                            | distill, wash
                            ┌───────────────┘
            anisole, methylene   p-methoxyacetophenone
            chloride, by-products   ──────────────────────

3. (a) $CH_3CH_2CH_2Br + AlCl_3 \rightarrow CH_3CH_2\overset{\oplus}{C}H_2(AlCl_3Br)^- \xrightarrow{\text{rearr.}}$

$CH_3\overset{\oplus}{C}H-CH_3(AlCl_3Br)^- \xrightarrow{\bigcirc}$ $\overset{(CH_3)_2CH \quad H}{\underset{}{\underset{\bigcirc}{+}}}$ $+ AlCl_3Br^- \rightarrow$

$\underset{\bigcirc}{CH_3CHCH_3}$ $+ AlCl_3 \pm HBr$

(b)

$\bigcirc$ $\xrightarrow[\text{AlCl}_3]{\text{CH}_3\text{CH}_2\overset{\overset{\text{O}}{\|}}{\text{C}}\text{-Cl}}$ $\bigcirc\overset{\overset{\text{O}}{\|}}{\text{C}}\text{-CH}_2\text{CH}_3$ $\xrightarrow{\text{Zn(Hg), HCl}}$ $\bigcirc\text{CH}_2\text{CH}_2\text{CH}_3$

(or use Wolff-Kischner reduction, etc)

4.

$\text{CH}_2=\text{CH}_2 \xrightarrow{\text{O}_2} \text{CH}_3\text{CHO} \xrightarrow[\text{catalysts}]{\text{O}_2} (\text{CH}_3\text{CO})_2\text{O} \xrightarrow{\bigcirc \ \text{AlCl}_3} \bigcirc\text{COCH}_3$

aceto-
phenone

$\bigcirc \left[\begin{array}{l} \xrightarrow{\text{Br}_2, \text{ Fe}} \bigcirc\text{Br} \xrightarrow[\text{NaOH}]{\text{MeOH}} \bigcirc\text{OCH}_3 \xrightarrow[\text{AlCl}_3]{\text{Ac}_2\text{O}} \bigcirc\begin{array}{c}\text{OCH}_3\\ \\ \text{COCH}_3\end{array} \\ \\ \xrightarrow[\text{AlCl}_3]{\text{CCl}_4} \text{Ph}_2\text{CCl}_2 \xrightarrow{\text{H}_2\text{O}} \text{PhCOPh} \end{array} \right.$

p-methoxy-
acetophenone

benzophenone

There are many other possible routes.

EXPERIMENT 28

Time: 1 period

Prelab calculations: 3.1 ml conc. nitric acid, 4.7 ml acetic anhydride, 26.6 ml toluene, 30.6 ml ethylbenzene, 34.9 ml isopropylbenzene, 38.7 ml t-butylbenzene.

Chemicals per 10 students:

acetic anhydride	65 g (60 ml)
t-butylbenzene	100 g (115 ml)
calcium chloride, anh.	20 g
ethylbenzene	80 g (92 ml)
isopropylbenzene (cumene)	90 g (105 ml)
nitric acid, conc.	60 g (42 ml)
toluene	70 g (80 ml)
2M sodium carbonate	500 ml (106 g Na$_2$CO$_3$)

Special equipment: Gas chromatograph(s)

Comments: The quantities specified above are based on an equal distribution of the four arenes. The gas chromato-graphy can be carried out on a silicone column at about 170°; we have used an SE-30 stationary phase on Aeropak #30 with satisfactory separation of the ortho and para (but not the meta) peaks. As stated in the Methodology, the product peaks generally are eluted in the order o, m, p following the alkylbenzene peak, and they are not hard to identify. Ortho:para ratios for toluene ethylbenzene, cumene, and t-butylbenzene are approximately 1.6, 0.8, 0.4, and 0.15, respectively, as compared to the statistically expected ratio of 2:1.

Topics for Report:

1. It should be apparent that a steric effect is operating, since the ortho:para ratio is, in all cases, less than 2:1 and since it decreases with an increase in the bulkiness of the alkyl group. The alkyl group, by steric crowding, destabilizes the aronium ion formed during ortho substitution. The alkyl groups in order of decreasing bulkiness are t-butyl > isopropyl > ethyl > methyl.

2. $HNO_3 + Ac_2O \rightarrow NO_2{}^+OAc^- + HOAc$

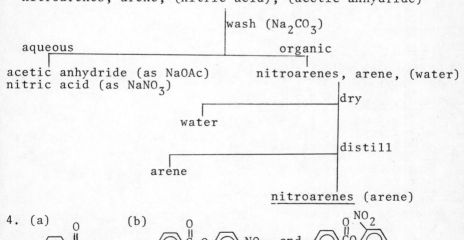

3.
```
        arene, nitric acid, acetic anhydride
                        │
                        │ reaction
                        │
    nitroarenes, arene, (nitric acid), (acetic anhydride)
                        │
                        │ wash (Na2CO3)
                        │
  aqueous               │               organic
            ┌───────────┴───────────────────┐
acetic anhydride (as NaOAc)      nitroarenes, arene, (water)
nitric acid (as NaNO3)                          │
                                                │ dry
                                    ┌───────────┤
                                    │           │
                                  water         │
                                                │ distill
                                    ┌───────────┤
                                    │           │
                                  arene         │
                                                │
                                    nitroarenes (arene)
```

4. (a)

O_2N—⬡—$\overset{\overset{O}{\|}}{C}OEt$

(and ortho)

(b) ⬡—$\overset{\overset{O}{\|}}{C}$—O—⬡—$NO_2$ and ⬡—$\overset{\overset{O}{\|}}{C}O$—⬡$NO_2$

55

(c)

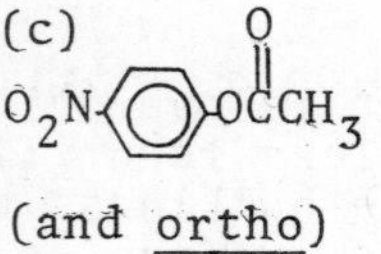

O_2N—⬡—$OCCH_3$
         $\overset{O}{\|}$

(and *ortho*)

(d) 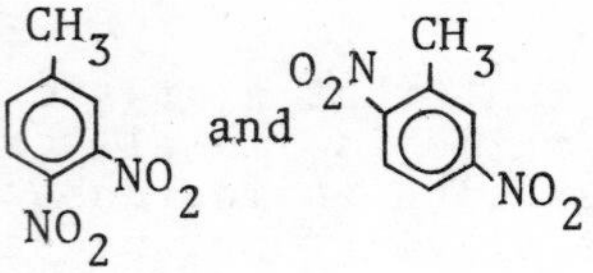 CH_3

⬡ with NO_2 and NO_2 groups

and O_2N—⬡—CH_3 with NO_2

(e) 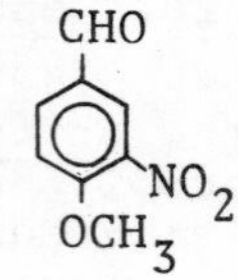 CHO

⬡ with NO_2 and OCH_3

MINILAB 12

Chemicals per 10 students:

naphthalene	15 g
nitric acid, conc.	21 g (15 ml)
sulfuric acid, conc.	28 g (15 ml)

The preparation should yield about 0.8 g of 1-nitronaphthalene, m.p. 58-9°.

Interpretation: $HNO_3 + 2H_2SO_4 \rightarrow NO_2^+ + 2HSO_4^- + H_3O^+$

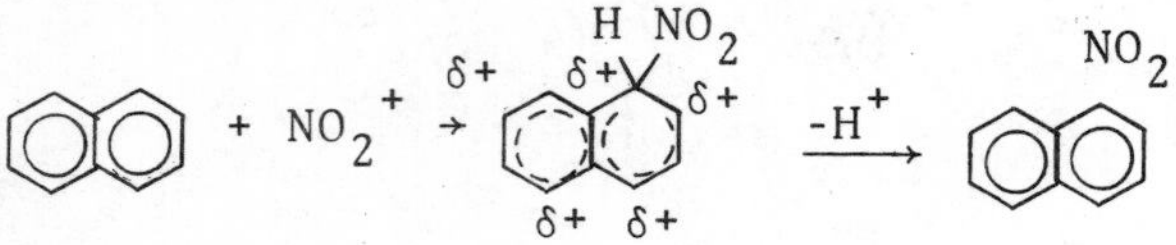

The aronium ion pictured for α-substitution has two resonance structures in which the aromatic structure of the second ring is preserved; the aronium ion for β-substitution has only one,

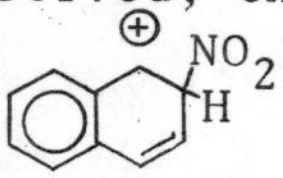

EXPERIMENT 29

Time: 1 period +

Prelab calculations: 1.48 g "cicelene", 6.32 g potassium permanganate.

Chemicals per 10 students:

anethole (Aldrich 11,787-0)	20 g (20 ml)
6M hydrochloric acid	250 ml (125 ml conc. HCl)
potassium permanganate	80 g
sodium bisulfite	90 g
1M sodium hydroxide	650 ml (26 g NaOH)
toluene	350 g (400 ml)
tricaprylmethylammonium chloride	7 g

Comments: "Cicelene" is anethole, p-$CH_3CH=CHC_6H_4OCH_3$, which yields p-methoxybenzoic acid in the oxidation reaction. "Tricaprylmethylammonium chloride", or the equivalent, can be obtained from Aldrich (Adogen 464), Tridom/Fluka (Aliquat 336), and Eastman Kodak (Methyltrioctylammonium chloride). Vigorous shaking is required--using a metal water bath is strongly recommended, since a large beaker might be broken accidentally. Both layers should be water-clear after the sodium bisulfite treatment (a little solid MnO_2 is allowable since it can be filtered off) and there

56

is usually an emulsion at the interface. Since some of the product may be trapped in the emulsion, the residue on the glass wool should be washed with some additional toluene. In the recrystallization, about 80-100 ml of water is needed per gram of p-methoxybenzoic acid, which crystallizes as nice white (or sl. yellow) needles, m.p. 187°. The product can be left in a dessicator jar overnight or dried in an oven before the melting point is obtained.

Analysis: The infrared bands needed to determine the side-chain structure are at about 960 and 830 cm^{-1}. In addition, there are asymmetric and symmetric C-O-C stretch bands at 1250 and 1030 cm^{-1} due to the methoxy group. A reproduction of the spectrum can be found in the back of this manual. The NMR spectrum of anethole (E.V. 5) is recorded in the Aldrich Library of NMR Spectra (4, 97C).

Topics for Report:

1. Oxidation of "cicelene" yields p-methoxybenzoic acid, indicating that the side chain is para to the methoxy group. From Table 2, it can be seen that the infrared band at about 830 cm^{-1} is due to the aromatic ring. Of the remaining bands below 1000 cm^{-1}, the 960 cm^{-1} band is the only one corresponding to a structure in Table 1, that being trans-RCH=CHR. Thus cicelene must have a 1,2-disubstituted double bond with the substituents trans to each other, and the structure illustrated is the only one possible. Of the other possible side chains, cis-CH_3C=CH-, $CH_3\dot{C}$=CH_2, CH_3CH=CH-, and cyclopropyl, the first three are eliminated by the fact that the infrared spectrum has no strong bands near 665-730, 890, or 910 and 990 cm^{-1}. Cyclopropane is eliminated because it was stated in the Situation that cicelene has an unsaturated side chain (see also E.V. 1).

2. (a) $3CH_3OC_6H_4CH$=$CHCH_3$ + $8KMnO_4$ → $3CH_3OC_6H_4COO^-K^+$ + $3CH_3COO^-K$ + $8MnO_2$ + $2H_2O$ + $2KOH$; (b) 0.010 mol x 8/3 x 158.0 = 4.21 g. Used 6.32 g, a 50% excess. (Further oxidation of potassium acetate to potassium carbonate is possible.)

3. $3ArCH$=$CHCH_3$ + $8Q^+MnO_4^-$ → $3ArCOO^-$ + $2H_2O$ + $2OH^-$ + $8Q^+$ + $8MnO_2$

toluene
water

$KMnO_4$ + Q^+ → $Q^+MnO_4^-$ + K^+

4. benzene $\xrightarrow[SO_3]{H_2SO_4}$ $PhSO_3H$ $\xrightarrow{NaOH, heat; HCl}$ $PhOH$ $\xrightarrow{CH_3I, NaOH}$

57

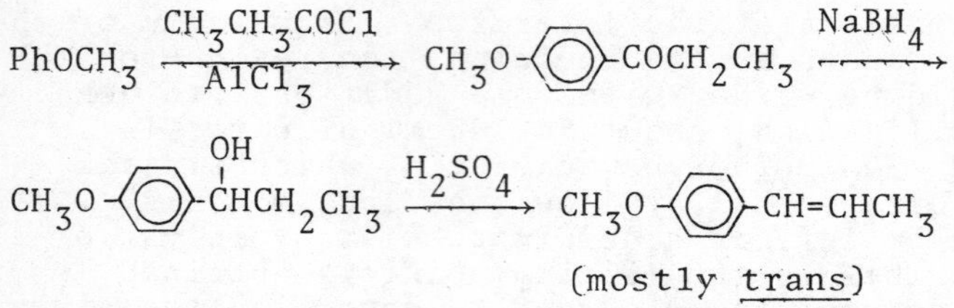

PhOCH$_3$ $\xrightarrow[\text{AlCl}_3]{\text{CH}_3\text{CH}_2\text{COCl}}$ CH$_3$O-◯-COCH$_2$CH$_3$ $\xrightarrow{\text{NaBH}_4}$

$\overset{\text{OH}}{\underset{}{}}$

CH$_3$O-◯-CHCH$_2$CH$_3$ $\xrightarrow{\text{H}_2\text{SO}_4}$ CH$_3$O-◯-CH=CHCH$_3$

(mostly <u>trans</u>)

(Alternatively, use CH$_3$OC$_6$H$_4$MgBr + CH$_2$=CHCH$_2$Br →

CH$_3$OC$_6$H$_4$CH$_2$CH=CH$_2$, isomerize with KOH)

5. Elemental analysis will give the empirical formula
(C$_6$H$_{12}$O$_3$); chemical tests and IR can show the presence of
a phenolic OH, alcoholic OH, and unsaturation; cleavage
with HI should indicate the presence of one methoxy group.
IR can show that the methoxy group is on the aromatic ring
and indicate 1,2,4-trisubstitution. At this point it
should be possible to deduce that coniferin is a methoxy-
phenol with a C$_3$H$_5$O side chain, and IR analysis (or NMR)
can be used to elucidate the structure of the side chain.
A student who is familiar with the structures of natural
products should recognize the relationship between con-
iferyl alcohol and vanillin, to which the former can be de-
graded by ozonolysis, showing the relative positions of the
substituents. Conferyl alcohol can be synthesized by an
aldol reaction of vanillin, followed by reduction:

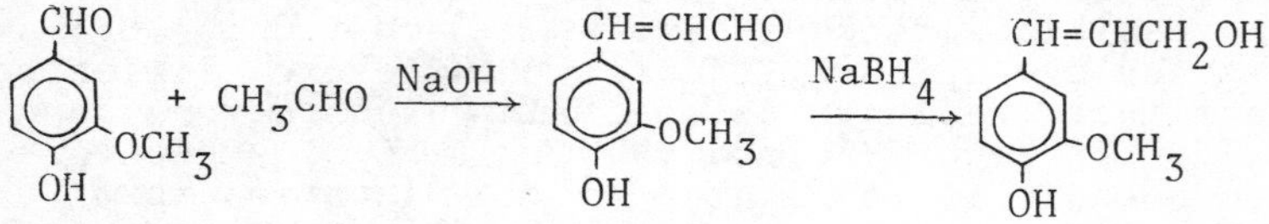

◯ CHO + CH$_3$CHO $\xrightarrow{\text{NaOH}}$ ◯ CH=CHCHO $\xrightarrow{\text{NaBH}_4}$ ◯ CH=CHCH$_2$OH
OCH$_3$ OCH$_3$ OCH$_3$
OH   OH   OH

6. anethole, potassium permanganate, toluene, water, Q$^+$

| react

<u>p</u>-methoxybenzoic acid (salt), OAc$^-$, OH$^-$, MnO$_2$, K$^+$; MnO$_4^-$,
Q$^+$, water, (anethole)

| sodium bisulfite,
| HCl

aqueous   organic

MnO$_2$ and MnO$_4^-$ (as Mn^{2+}),   <u>p</u>-methoxybenzoic acid,
K$^+$, OH$^-$ (as H$_2$O), OAc$^-$ toluene, Q$^+$, (water),
(as HOAc), Q$^+$ (anethole), (acetic acid)

| extract (NaOH)

(Continued on next page)

58

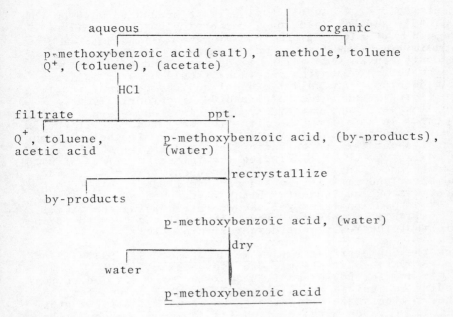

```
                    |
  aqueous           |        organic
 ┌─────────────────────────────────────┐
 p-methoxybenzoic acid (salt),  anethole, toluene
 Q⁺, (toluene), (acetate)
           │
           │HCl
           │
filtrate   │             ppt.
┌──────────┴──────────────────────┐
Q⁺, toluene,      p-methoxybenzoic acid, (by-products),
acetic acid       (water)
                          │
          ┌───────────────┤recrystallize
          │               │
     by-products          │
                          │
                 p-methoxybenzoic acid, (water)
                          │
          ┌───────────────┤dry
          │               │
       water              │
                          │
                 p-methoxybenzoic acid
                 ─────────────────────
```

EXPERIMENT 30

Time: 1 period

Prelab calculation: 0.040M x 0.025 ℓ x 215 g/mol = 0.215 g
AdBr

Chemicals and supplies per 10 students (or pairs)[*]:

1-bromoadamantane (Aldrich 10,922-3)	3 g
*bromothymol blue indicator	25 ml (0.010 g BTB)
ethanol (95% or absolute)	1.6 kg (20 ℓ)
methanol	515 g (650 ml)
isopropyl alcohol	390 g (500 ml)
phosphate buffer	300 ml
*2 x 10⁻⁵M sodium hydroxide	1 ℓ (0.08 g NaOH)
burets, 25 ml	10
volumetric flasks, 25 ml	10
volumetric pipets 1 ml	10

Solution preparations: Bromothymol blue solution - dissolve
0.01 g BTB in 8 ml of 2 x 10⁻⁵M NaOH and make up to 25 ml
with water. Sodium hydroxide solution - make up 100 ml of
0.2M NaOH using 0.8 g NaOH, dilute 10 ml of this to 1 liter.

Comments: It is important to keep the 1-bromoadamantane dry
and to avoid moisture in making up the 0.04M solution. The
exact concentration of this solution is not important since
t_α is independent of concentration. Likewise, the NaOH
solution need not be standardized. If desired, 95% ethanol
can be used to make up the ethanol solutions, in which case
*The adamantyl bromide must be pure and dry or the experiment will not
work properly; it can be purified by vacuum sublimation if necessary.

59

about 42.1, 47.4, and 52.6 ml will be needed to make up the 40%, 45%, and 50% solutions, respectively. **The buffer is prepared by dissolving lg each of disodium hydrogen phosphate and potassium dihydrogen phosphate in 300 ml of deionized water.** Students can work in pairs, if desired, and combine their glassware. Most of the reactions should be over in less than 50 minutes, with 40% ethanol and 50% methanol reacting the fastest, and 50% ethanol and 40% isopropyl alcohol the slowest. Duplicate runs can be started before all of the solutions from the first run have reacted, if enough flasks or other containers are available. If the solutions are of about the concentrations designated, it should take approximately 20 ml (including the initial 2 ml) of NaOH to titrate the V_∞ mixtures; α is then about $2/20 = 0.1$. The indicator endpoint is light green and is difficult to pin-point without an indicator blank - students often wait un-til the solution turns yellow before recording the time. If available, a thermostatted bath (25^0) can be used to in-sure that the reaction times are reproducible.

From the Schleyer paper (J. Am. Chem. Soc. 92, 5977 (1970)), the 25^0 rate constants are, for 40% ethanol, 1.21×10^{-4} sec^{-1}; and for 50% ethanol, 2.86×10^{-5} sec^{-1}. The corres-ponding Y_{Ad} values are 2.38 and 1.76. Y_{Ad} values for the other solvents are estimated to be about 2.1 for 45% etha-nol, 2.2 for 50% methanol, and 1.7 for 40% isopropyl alco-hol. In this experiment, extra water is added to the sol-vents with the NaOH, so their ionizing powers should be somewhat greater than expected (for example, 2 ml of NaOH and 1 ml of AdBr/EtOH makes the "50% ethanol" solution about 48% ethanol instead).

Topics for Report:

1. The order should be 40% ethanol, 50% methanol > 45% etha-nol > 50% ethanol, 40% isopropyl alcohol. The first and last two solvent mixtures are very similar in ionizing pow-er. From these results it can be concluded that methanol has the greatest ionizing ability, followed by ethanol and isopropyl alcohol.

2.

(a)

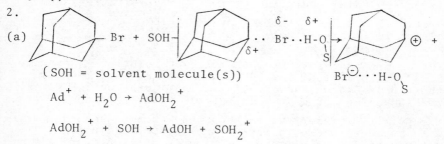

(SOH = solvent molecule(s))

$$Ad^+ + H_2O \rightarrow AdOH_2^+$$

$$AdOH_2^+ + SOH \rightarrow AdOH + SOH_2^+$$

(b) Adding more water to an alcohol/water mixture increases the ionizing power of the solvent, as seen by the fact that the hydrolysis rate increases in going from 50% ethanol to

40% ethanol. This is because water has a higher dielectric constant and a greater hydrogen-bonding potential than an alcohol, and is therefore more able to assist the removal of bromide ion in the rate-determining step.

(c) Methanol has the greatest ionizing power, followed by ethanol and isopropyl alcohol. This is to be expected on the basis of solvent polarity, methanol being the most polar (and thus most able to solvate the leaving group) and 2-propanol the least.

3. Add a weighed quantity of dry AdBr directly to the prepared solvent mixture. Withdraw aliquots at intervals, quench them in a comparatively non-reactive solvent (acetone works well), and titrate them to the BTB endpoint with NaOH. Alternatively, the reaction rate can be followed by recording the pH or conductance of the reaction mixture as a function of time.

4.

$$CH_3-\underset{\underset{CH_3}{|}}{\overset{\overset{Cl}{|}}{C}}-CH_3 \xrightarrow{SOH} CH_3\overset{\oplus}{\underset{\underset{CH_3}{|}}{C}}-CH_3 + Cl^- \text{ (solvated)}$$

$$CH_3-\overset{\oplus}{\underset{\underset{CH_3}{|}}{C}}-CH_3 + SOH \rightarrow CH_3-\underset{\underset{CH_3}{|}}{C}=CH_2 + SOH_2^+$$

Elimination from the adamantyl carbonium ion would yield a double bond at the bridgehead, which is forbidden because of bond-angle strain.

5. Tricyclo $[3.3.1.1^{3,7}]$ decane.

EXPERIMENT 31

Time: 1 period

Prelab calculations: 10.0 g/0.75 = 13.3 g menthone (theoretical) 13.3 x $\frac{156}{154}$ = 13.5 g menthol needed. Scaling factor = 13.5/90 = 0.150

Suggested procedure: Place 17.7 g (60 mmol) potassium dichromate in a 250 ml Erlenmeyer flask and mix in a solution of 8.1 ml concentrated sulfuric acid in 90 ml of water. To this mixture add 13.5 g of menthol, in 3-4 portions, with shaking and stirring (stirring rod okay). Monitor the temperature; heat the mixture gently (steam bath recommended) if it does not reach 55°. When the temperature falls, cool the mixture to room temperature and add about 100 ml of ethyl ether. Mix thoroughly and transfer the mixture to a 250 ml separatory funnel, using a

little ether in the transfer. Drain off the aqueous layer
and wash the ether layer with three 30-ml portions of 5%
sodium hydroxide solution. Dry the solution with 3-4 g of
magnesium sulfate and evaporate the ether from a steam bath.
Purify the residue by semimicro simple distillation, collecting the menthone at 204-7^a.

Chemicals per 10 students:

ethyl ether	1.1 kg (1.5 ℓ)
menthol (Aldrich M277-2)	175 g
potassium dichromate	225 g
sulfuric acid, conc.	200 g (110 ml)
5% sodium hydroxide	1.2 ℓ (63 g NaOH)

Comments: One statement in the O.S. procedure, "The oil is
....washed with three 200-cc portions of 5 per cent sodium
hydroxide solution..." can lead to confusion. It is, of
course, the ether in which the oil was dissolved that is
washed with the NaOH. Although the O.S. procedure does not
say that the ether layer should be dried, students should
recognize that this is advisable. The product can be purified by vacuum distillation to make the preparation a bit
more challenging, but simple distillation works nearly as
well. Optically active (-)-menthol for E.V. 1 can be obtained from Aldrich, Eastman Kodak, and other suppliers:
it is about 50% more expensive than the racemate. The I.R.
spectra of natural and synthetic menthol (E.V. 2) are recorded in the Aldrich collection, 2nd edition, 94D and 94E.

Topics for Report:

1. $\dfrac{13.5 \text{ g}}{156.3}$ = 0.0864 mol menthol, requiring 0.0288 mol of
potassium dichromate (8.47 g). Actually used 0.0600 mol
(17.7 g) of dichromate, a 108% excess.

2.   No. stereoisomers = 2^3 = 8

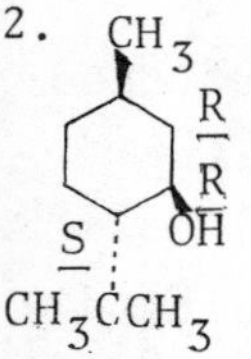

3.

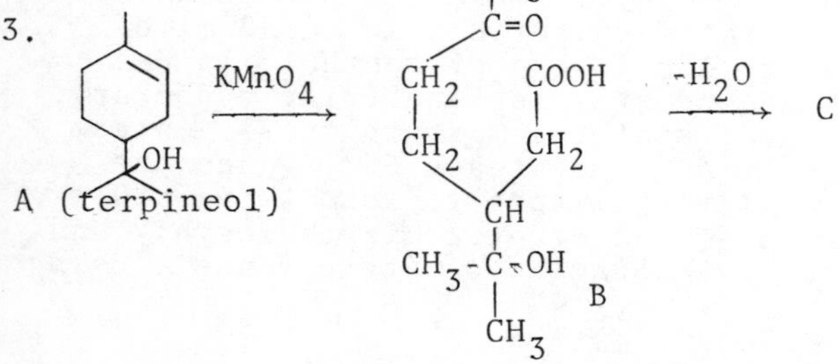

Oxidation cleaves the terpineol ring at the double bond.

The resulting γ-hydroxycarboxylic acid dehydrates to yield the lactone.

4.   menthol, potassium dichromate,
                    sulfuric acid, water

                    | react

menthone, Cr^{3+}, K^+, SO_4^{2-}, $Cr_2O_7^{2-}$, H^+, water, (menthol)

                    | extract (ether)
                    | wash (NaOH)

   aqueous   organic

Cr salts, K^+, SO_4^{2-},   menthone, ether,
H^+ (as H_2O)   (menthol) (water)

                                            | dry

                    water

                                            | evaporate

                ether

                                            | distill

            forerun   residue

            ether   __menthone__ menthol

5.

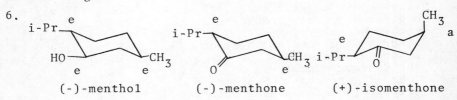

Enolization, followed by addition of H^+ to the other side of the ring.

6.

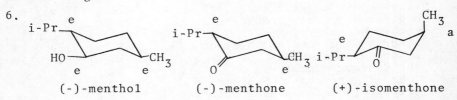

        (-)-menthol   (-)-menthone   (+)-isomenthone

MINILAB 13

__Chemicals per 10 students:__

1-butanol	1½ g (2 ml)
2-butanol	1½ g (2 ml)
3M hydrochloric acid	6 ml (1½ ml conc. HCl)

methanol	1½ g (2 ml)
2-methyl-2-propanol	
(t-butyl alcohol)	1½ g (2 ml)
0.05% potassium permanganate	130 ml (0.065 g KMnO$_4$)
sulfuric acid, conc.	4 g (2 ml)

Comments: 2-Butanol decolorizes the permanganate solution completely after 10 minutes in neutral solution, 1-butanol partially reacts, and the others show no reaction. After adding dilute HCl, 1-butanol reacts completely and methanol nearly completely after 10 minutes. The tertiary alcohol shows only a slight reaction at this point, but it reacts almost completely after the sulfuric acid is added.

Interpretation: 2-butanol > 1-butanol > methanol > t-butyl alcohol (2^0 > 1^0 > CH$_3$ > 3^0)

$$3CH_3CH_2\overset{\overset{OH}{|}}{C}H\text{-}CH_3 + 2MnO_4^- \rightarrow 3CH_3CH_2\overset{\overset{O}{\|}}{C}\text{-}CH_3 + 2MnO_2 + 2H_2O + 2OH^-$$

$$3CH_3CH_2CH_2CH_2OH + 4MnO_4^- \rightarrow$$
$$3CH_3CH_2CH_2COO^- + 4MnO_2 + 2H_2O + 3OH^- \text{ (neutral solution)}$$

$$5CH_3OH + 4MnO_4^- + 12H^+ \rightarrow 5HCOOH + 4Mn^{2+} + 11H_2O \text{ (acid solution)}$$

$$5CH_3\overset{\overset{OH}{|}}{\underset{\underset{CH_3}{|}}{C}}CH_3 + 6MnO_4^- + 18H^+ \rightarrow 5CH_3\overset{\overset{O}{\|}}{C}\text{-}CH_3 + 5HCOOH + 6Mn^{2+} + 14H_2O$$

t-Butyl alcohol dehydrates to isobutylene in the presence of strong acid, and the alkene is oxidized further.

EXPERIMENT 32

Time: 1-1½ periods

Chemicals per 10 students:

ethyl bromide (Aldrich 12,405-2)	70 g (48 ml)
magnesium sulfate, anhydrous	35 g
methylene chloride	1.1 kg (850 ml)
saturated aq. sodium chloride	325 ml (100 g NaCl)
2M sodium hydroxide	650 ml (52 g NaOH)
90% aqueous phenol	35 g
(or 30 g crystalline phenol)	
tricaprylmethylammonium chloride	13 g

Comments: This is the only experiment that cannot be readily performed without a magnetic or mechanical stirrer--it is hardly practical to have students shake the apparatus continually throughout a two-hour reflux period. Most students should be able to complete the experiment in a single four-hour period. Otherwise, a good stopping place is just after drying--if feasible, students can leave their methylene chloride solutions under a hood to evaporate overnight and carry out the distillation during the next lab period. If students are provided with ordinary 19/22 labkits, they can use the distillation column as a second reflux condenser. The reaction should be complete in two hours with efficient stirring--if time permits, the reflux period can be extended an hour or so. The workup is routine, though some foaming may be encountered during the evaporation of methylene chloride. Considerable product can be lost in the distillation, so the smallest available short-path distillation apparatus should be used. Sometimes a dense white vapor forms in the distillation flask after most of the phenetole has distilled. Although this doesn't appear to represent a hazard, the distillation should be stopped at this point (the residue boils over 180^0). Because of the small scale of the experiment, quite a bit of care is needed to get yields much over 75%.

The general theory and practice of ether synthesis with phase-transfer catalysis are discussed in the article by A. McKillop, J.-C.Fiaud, and R.P. Hug in Tetrahedron, 30, 1379 (1974). The authors used 1 m Carbowax 20M columns for gas chromatographic analysis of the products (E.V. 1). Spectra are reproduced in the Sadtler Standard Spectra (IR, prism #192, grating #4827) and in the JEOL High Resolution NMR Spectra, #102.

Topics for Report:

1. Q^+X^- + NaOH → Q^+OH^- + NaCl (aqueous phase) (X = Cl or Br)

   Q^+OH^- + PhOH → Q^+OPh^- + H_2O (aqueous phase)

   Q^+OPh^- + CH_3CH_2Br → $PhOCH_2CH_3$ + Q^+Br^- (organic phase)

2.   phenol, ethyl bromide, sodium hydroxide,
             methylene chloride, water, Q^+Cl^-

                                  | reflux,
                                  | separate

aqueous   organic

extract (CH_2Cl_2)

Na^+, OH^-, Br^-, (phenetole) —→   phenetole, ethyl bromide
Q^+, Cl^-, phenol   (xs), ethanol*, methylene
                                  chloride, (phenol), (Q^+)
                                      (water)

*by-product from
hydrolysis of ethyl   wash (NaOH,
bromide.   NaCl)
                    Q^+, phenol
                    (as salt)   (Continued on next page)

65

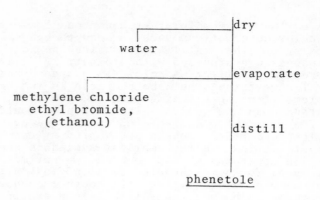

phenetole

3. (a) $Na^+OPh^- + CH_3CH_2Cl \rightarrow PhOCH_2CH_3 + NaCl$

$Na^+OPh^- + CH_3$⟨⟩$SO_2OCH_2CH_3 \rightarrow PhOCH_2CH_3 + CH_3$⟨⟩$SO_2O^-Na^+$

$2Na^+OPh^- + (CH_3CH_2)_2SO_4 \rightarrow 2PhOCH_2CH_3 + Na_2SO_4$

$3Na^+OPh^- + (CH_3CH_2)_3PO_4 \rightarrow 3PhOCH_2CH_3 + Na_3PO_4$

(b) $-Cl;$ CH_3⟨⟩$SO_2O-;$ $CH_3CH_2OSO_2O-;$ $(CH_3CH_2O)_2\overset{O}{\underset{}{P}}O-$

4. 1. $CH_2CH_2CH_2Br + PhOH \xrightarrow{NaOH, \ CH_2Cl_2} PhOCH_2CH_2CH_3$

2. $(CH_3)_3COH \xrightarrow{Na} (CH_3)_3CO^-Na^+ \xrightarrow{CH_3CH_2Br} (CH_3)_3COCH_2CH_3$

3. [structure with CH_3, OH, OH] $+ CH_2Br_2 \xrightarrow{NaOH}_{PTC}$ [structure with CH_3, O, $O-CH_2$] (see E.V. 2)

4. $PhOH + PhCH_2Br \xrightarrow{NaOH}_{PTC} PhOCH_2Ph$

5. [structure with OH, OH] $+ CH_3I \xrightarrow{NaOH}_{PTC}$ [structure with OCH_3, OCH_3]

(Phase-transfer catalysts are optional. Alkylating agents
can be synthesized from the corresponding alcohols by ROH
+ HBr, ROH + H_2SO_4, etc.)

EXPERIMENT 33

Time: 1 period +

Prelab Exercise: The following procedure can yield in the
vicinity of 90% vanillyl alcohol. The parameters can be
varied within rather broad limits to give at least a finite
yield of product (see Comments below).

Dissolve 7.60 g of vanillin (50 mmol) in 55 ml of 1M NaOH
in a beaker, and dissolve 1.0 g (26 mmol) of sodium borohydride in 5-10 ml of 1M NaOH in a 250 ml Erlenmeyer flask.
Cool both solutions to 15° and add the vanillin solution to
the borohydride solution in portions, monitoring the temperature to see that it does not rise above 25°. Let the
reaction mixture stand at room temperature for 30-40 minutes with frequent swirling or magnetic stirring, then
acidify it with 6M HCl to a pH of 6 or less (test for complete precipitation). Collect the vanillyl alcohol by
vacuum filtration (wash with cold water) and recrystallize
it from boiling water. Dry the product thoroughly in a
dessicator before obtaining the melting point.

Chemicals per 10 students:

6M hydrochloric acid	180 ml (90 ml conc. HCl)
sodium borohydride	12 g
sodium hydroxide	40 g (or 1 ℓ 1M NaOH)
vanillin (Aldrich V110-4)	100 g
(Optional: 80 ml each of methanol, ethanol, 2-propanol)	

Comments: A number of options are possible for this experi-
ment--you can have students work individually and compete
with one another to see who can come up with the best yield
of pure product, or students can cooperate with each other
in small groups to see which group can develop the best
procedure. With a suitable procedure, the average student
should be able to complete the experiment easily in a short
laboratory period; with an ill-conceived one, a student may
struggle through several periods. The amount of guidance
provided is up to the instructor--wasted time and frustra-
tion can be considerably reduced by going over each student's procedure and, by asking pointed questions, leading
students away from pitfalls. (On the other hand, frustra-
tion is certainly a part of chemical research and you may
prefer to let them learn by their mistakes.)

After carefully reading the Methodology, students should
recognize that vanillin, being a phenol, should be converted to a salt to keep it from decomposing part of the reducing agent. This should lead them to the conclusion that
dilute aqueous sodium hydroxide would be the most convenient solvent for the reaction, and that over 50 mmol of

NaOH should be used to neutralize the vanillin and keep the reaction mixture alkaline. Alcoholic solvents can be used with an excess of the borohydride, but the workup is more difficult and the yields generally lower. Some students may come up with the reasonable idea of converting vanillin to the salt by dissolving it in an equivalent amount of concentrated NaOH and then adding alcohol--unfortunately, the salt is insoluble in alcohol. When a procedure like the one outlined above is used, the pH of the reaction mixture should be checked to see that it is over 10, and plenty of time (ice cooling) should be allowed to insure complete precipitation of vanillyl alcohol, both from the initial reaction mixture and from the recrystallization solvent. There is considerable foaming when the reaction mixture is acidified, and most of the product may end up in the foam. Vanillyl alcohol has also been purified by recrystallization from a 1:3 mixture of ethyl acetate and pentane (see Chem. Abstracts 64, 9620e (1966)). The product has a much milder vanilla odor than vanillin itself, and it is quite possible that the odor is due to vanillin as an impurity. The IR spectrum of vanillyl alcohol (E.V. 1) is reproduced in the Aldrich Library of Infrared Spectra, 2nd ed., p. 612F.

Topics for Report:

1. In order for students to draw meaningful conclusions, they should have the results from as many groups as possible, and the workup procedures should be comparable.

2. $Na^+BH_4^- + 3H_2O + HCl \rightarrow 4H_2 + H_3BO_3 + NaCl$

3.   vanillin, sodium borohydride, NaOH, water

                                 | reaction

        vanillyl alcohol, sodium borohydride (xs),
                NaOH, (vanillin), water

                                 | add HCl,
                                 | filter

    filtrate   precipitate

Na^+, BH_4^- (as boric acid),   vanillyl alcohol, (vanillin),
OH^- (as H_2O), H^+, Cl^-, (H_2) (water)

                                        | recrystallize

        vanillin   vanillyl alcohol, (water)

                                        | dry

        water   vanillyl alcohol

4. If borohydride is in excess during the reaction of a

68

base-sensitive material, the excess borohydride can pro-
mote unwanted side reactions. Therefore the substrate
should be kept in excess, insofar as possible, by adding
the borohydride solution <u>to</u> the substrate.

5.
$$\text{Ar-}\overset{\overset{\displaystyle O}{\|}}{\text{C}}\text{-H} + BH_4^- \rightarrow \text{Ar-}\overset{\overset{\displaystyle O-BH_3^-}{|}}{\underset{\underset{\displaystyle H}{|}}{\text{C}}}\text{-H} \xrightarrow{\text{ArCHO}} (ArCH_2O)_2BH_2^- \xrightarrow{\text{ArCHO}}$$

$$(ArCH_2O)_3BH^- \xrightarrow{\text{ArCHO}} (ArCH_2O)_4B^-$$

$$Ar = $$

(The actual mechanism may be more complex than this)

EXPERIMENT 34

<u>Time</u>: 2-3 periods

<u>Prelab calculations</u>:   15.92 g (15.3 ml) benzaldehyde

<u>Chemicals per 10 students</u>:

80% acetic acid	450 ml (390 g HOAc)
ammonium nitrate	100 g
benzaldehyde	200 g (190 ml)
carbon, decolorizing	12 g
copper(II) acetate	2 g
95% ethanol	1.3 kg (1.7 ℓ)
filtering aid (Celite, etc.)	6 g
hydrochloric acid, conc.	225 g (190 ml)
methanol	200 g (250 ml)
6M potassium hydroxide	500 ml (200 g 85% KOH pellets)
2M sodium cyanide	125 ml (12.25 g NaCN)

<u>Comments</u>: The benzaldehyde should be freshly opened or re-
cently distilled for good results. Students should wear
rubber gloves while handling cyanide solutions. The reaction mixture from part A should be transferred, while
still hot, to an Erlenmeyer flask for cooling. A burner
works best for the Part B reflux, though a mantle can also
be used. About 4 ml of 95% ethanol per gram of benzil is
sufficient for the recrystallization. A nice display of
crystallization from a supersaturated solution can be observed if the solution is allowed to cool slowly to room
temperature, then seeded with a crystal of pure benzil;
however, the product will be purer if crystallization is
induced by scratching the sides of the flask while the
solution is cooling. The benzil does not have to be completely dry before beginning step C, but it should be dry
enough to get a reasonably accurate mass. If an oven is
used for drying, its temperature should be below 80° so
that the benzil does not melt. Students should be able to
complete Part A and the reaction phase of Part B in one
four-hour period.

The reaction mixture in Part C is initially a purple-black

color but turns brown during the reflux period. It is best
to filter the hot Norit-treated solution through a very
fast fluted filter paper first, then re-filter it through a
medium paper like Whatman #1 (fluted) to remove all of the
carbon. The crude benzilic acid requires a considerable
volume of hot water for recrystallization, and tends to
yield a gummy mass that is difficult to get into solution.
The recrystallization may, if desired, be omitted and the
students' crude benzilic acid dried and weighed for the
yield calculations. Al alternative purification method is
described in Vogel (Bib-B4). Yields of 80-90% for each
step are attainable, though the last recrystallization may
reduce the overall yield considerably.

Topics for Report:

1. $PhCOOH + CN^{\ominus} \rightarrow PhCOO^{\ominus} + HCN$
Benzoic acid is a much stronger acid than HCN, so it would
protonate cyanide ion and inactivate it as a catalyst.

2. (a) $Ph\overset{O}{\overset{\|}{C}}\text{-H} + CN^{\ominus} \rightarrow Ph\text{-}\overset{O^{\ominus}}{\overset{|}{\underset{CN}{C}}}\text{-H} \xrightarrow[\text{transfer}]{\text{proton}} Ph\text{-}\overset{OH}{\underset{CN}{C}}{:}^{\ominus} \xrightarrow[\text{addition}]{PhCHO \atop \text{nucl.}} Ph\text{-}\overset{\ominus O}{\underset{H}{C}}\text{-}\overset{OH}{\underset{CN}{C}}\text{-Ph}$

nucleophilic
addition

$\xrightarrow[\text{transfer}]{\text{proton}} Ph\text{-}\overset{HO}{\underset{H}{C}}\text{-}\overset{O^{\ominus}}{\underset{CN}{C}}\text{-Ph} \rightarrow Ph\text{-}\overset{HO}{\underset{H}{C}}\text{-}\overset{O}{\overset{\|}{C}}\text{-Ph}$ (the last step resembles the first one in reverse)

(b) $Ph\text{-}\overset{O}{\overset{\|}{C}}\text{-}\overset{O}{\overset{\|}{C}}\text{-Ph} \xrightarrow[\text{nucl.} \atop \text{addition}]{OH^-} Ph\text{-}\overset{O}{\overset{\|}{C}}\text{-}\overset{O^{\ominus}}{\underset{Ph}{C}}\text{-OH} \xrightarrow[\substack{\text{rearrangement} \\ \text{(via "nucleophilic} \\ \text{addition" of Ph)}}]{} Ph\text{-}\overset{\ominus O}{\underset{Ph}{C}}\text{-}\overset{O}{\overset{\|}{C}}\text{-OH} \xrightarrow[\text{transfer}]{\text{proton}} Ph\text{-}\overset{HO}{\underset{Ph}{C}}\text{-}\overset{O}{\overset{\|}{C}}\text{-}O^{\ominus}$

3.

furfural

$\xrightarrow{NaCN}$ $\xrightarrow[Cu^{2+}]{NH_4NO_3}$

furil

$\xrightarrow{KOH} \xrightarrow{HCl} 1$

4.   benzaldehyde, sodium cyanide, ethanol, water

reflux

benzoin, NaCN, ethanol, water, (benzaldehyde)
                                    (Continued on next page)

70

```
                                    │ filter, wash
         filtrate                   │
                                    │
    NaCN, ethanol         benzoin
  water, benzaldehyde
                                    │ Cu(OAc)₂, NH₄NO₃,
                                    │ reflux

    benzil, copper(II) acetate, ammonium nitrate (xs),
       acetic acid, water, nitrogen, (benzoin)

                                    │ ice water,
                                    │ filter
         filtrate
           │
  Cu²⁺, OAc⁻, NH₄⁺,        benzil, (benzoin),
  NO₃⁻, H⁺, (N₂), HOAc
                                    │ recrystallize (EtOH)
         │
    benzoin                   benzil
                                    │ KOH, EtOH,
                                    │ water,
                                    │ reflux

    benzilic acid (K⁺ salt), KOH, ethanol,
        water, (benzil), (by-products*)

                                    │ water, carbon,
  *by-products probably            │ filter
  include benzhydrol
                                            residue

                                    benzil, by-products,
                                    carbon

                                    │ add HCl,
                                    │ filter, wash

       filtrate
         │
  K⁺, OH⁻(as water)      benzilic acid, (benzil),
  H⁺, Cl⁻, ethanol,           (by-products)
  water
                                    │ recrystallize

       filtrate
         │
     benzil,            benzilic acid
     by-products
```

$Cu(OAc)_2$, NH_4NO_3

Cu^{2+}, OAc^-, NH_4^+, NO_3^-, H^+, (N_2), HOAc

benzilic acid (K^+ salt)

K^+, OH^- (as water), H^+, Cl^-, ethanol, water

MINILAB 14

Chemicals per 10 students:

benzil	13 g
95% ethanol	80-160 g (100-200 ml*)
o-phenylenediamine	7 g

*Use 200 ml if the product is to be recrystallized.

The melting point of 2,3-diphenylquinoxaline should be 125-6°. The yield is about 50%.

Reaction:

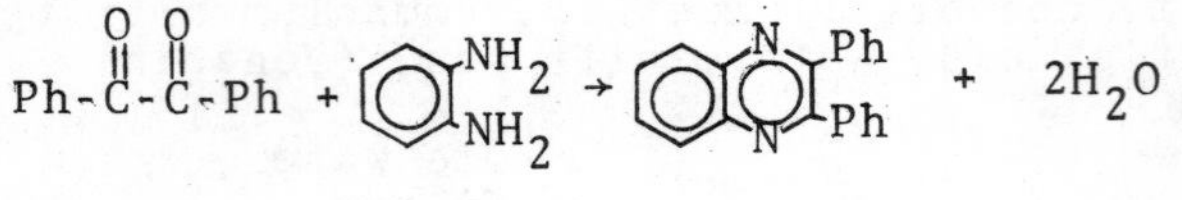

Mechanism:

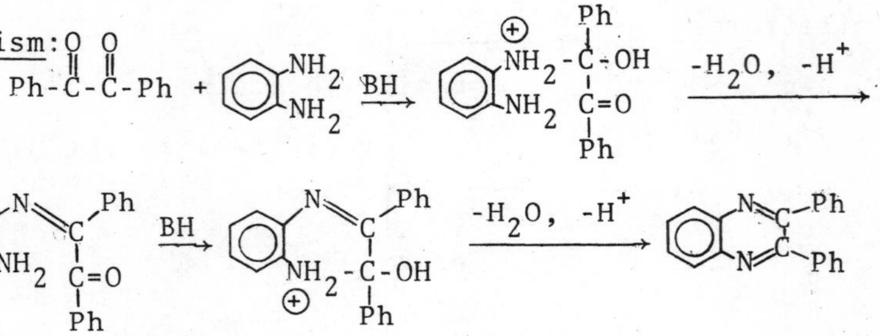

(BH = ROH or H_2O)

EXPERIMENT 35

Time: 1½-2 periods

Prelab calculations: 8.41 g (8.9 ml) cyclopentanone; 9.61 g (8.3 ml) furfural.

Chemicals per 10 students:

cyclopentanone	110 g (115 ml)
deuterochloroform (optional)	4 g (6 ml)
95% ethanol	1 kg (1.3 ℓ)
ethyl ether	850 g (1.2 ℓ)
furfural	125 g (110 ml)
magnesium sulfate, anhyd.	50 g
sodium chloride, satd. sol.	1.2 ℓ (375 g NaCl)
0.1M sodium hydroxide	1.2 ℓ (4.8 g NaOH)
tetrabutylammonium bromide	13 g

Comments: The by-product is 2,5-difurfurylidenecyclopentanone (Topic 1) resulting from condensation at both α-carbon atoms of cyclopentanone. Several grams of this product should be isolated. If desired, the experiment can be performed to prepare and identify the by-product only, in which case only the first sentence of the Separation and the last paragraph of the Purification and Analysis are required. This preparation can be completed easily in a short

lab period. If NMR spectrometers are not available, students can be provided with copies of the NMR spectrum reproduced in the back of this manual. (A Sadtler NMR spectrum of this by-product in Polysol-d is incorrectly labeled 2-furfurylidenecyclopentanone; it is spectrum #20 in
the 1975 supplement.)

The experiment can be carried out without the phase-transfer catalyst if the reaction mixture is shaken or stirred
for an hour or more after the furfural is added. The workup can be a little messy since the bis compound may con-
tinue to precipitate in the initial filtrate. It may be
necessary to re-filter this mixture and to add more ether
to replace any that evaporates during the reaction. 50-70
ml of saturated NaCl solution should be sufficient to wash
the combined ether extracts, and 3-4 g of magnesium sulfate
to dry them. The ether and unreacted starting materials
can be distilled off at atmospheric pressure if desired,
followed by vacuum distillation of 2-furfurylidenecyclo-
pentanone. The receiver for the final distillation should
not be cooled; otherwise the product may crystallize in the
vacuum adapter. Ordinarily it distills as a yellow liquid
that crystallizes on standing. The bis compound requires
about 35 ml of ethanol per gram for recrystallization; it
should crystallize as beautiful orange needles, m.p. 166-
168^0. Its NMR spectrum is discussed in Topic 1 below.

By analogy with other Claisen-Schmidt condensation products
(E.V. 1) one would expect 2-furfurylidenecyclopentanone
(and presumably the bis product as well) to have the E con-
figuration, with the furan ring trans to the carbonyl group.
However, my testers have not been able to photochemically
isomerize the product in order to establish this fact.
This might make a challenging research project for one of
your better students.

Topics for Report:

1. (a)   The NMR spectrum shows five sig-
                                 nals:
                                 a. 3.0δ, singlet, area = 2
                                 b. 6.5-6.6δ, complex, area = 1
                                 c. 6.7-6.8δ, complex, area = 1
                                 d. 7.2δ, singlet, area = 1
                                 e. 7.5δ, singlet, area = 1
2,5-difurfurylidenecyclopentanone

The upfield signal (a) is obviously due to the sp^3 cyclopentanone ring protons. The fact that it is a singlet
indicates that these protons are equivalent, so the ring
must be symmetrically substituted, presumably at the 2
and 5 positions. The downfield protons are in a chemical
shift range corresponding to aromatic and vinylic protons
and the signal pattern is consistent with a furfurlidene
group (three aromatic and one vinylic protons, none of them
equivalent). The ratio of furfurylidene protons to

73

alicyclic protons is 2:1, which is consistent with two furfurylidene groups attached to a single cyclopentane ring.
This product is consistent with the reaction mechanism,
which involves attack at the alpha-carbons of cyclopentanone. (The NMR spectrum of 2-furfurylidenecyclopentanone
shows a 3:2 ratio between the upfield and downfield signal areas, which is consistent with one furfurylidene
group (4 protons) substituted on a cyclopentanone ring with
6 protons. In addition the upfield pattern is no longer a
sharp singlet but a grouping of three complex signals, as
expected for the unsymmetrical substitution.) (b) See 1(a)
for assignments. Chemical shifts on the furfural ring
should be highest for the e-proton adjacent to oxygen, and
lowest for the b-proton. Assignment of the vinyl proton
signal is open to conjecture, but the 7.2δ singlet appears
most likely since the corresponding signal in the 2-fur-
furylidenecyclopentanone spectrum couples with the α-pro-
ton signal on the cyclopentanone ring.

2. (a)

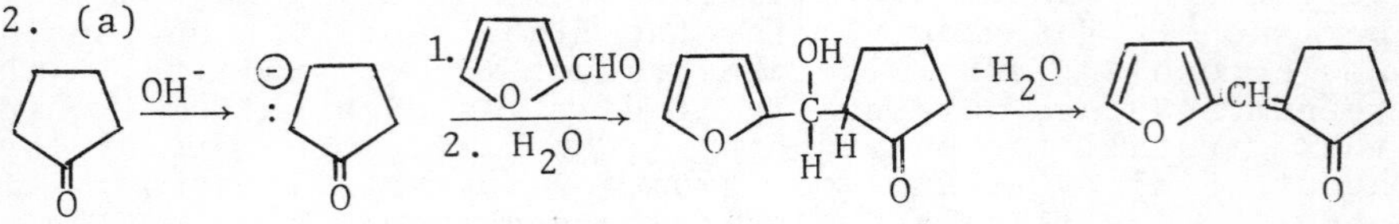

(b)

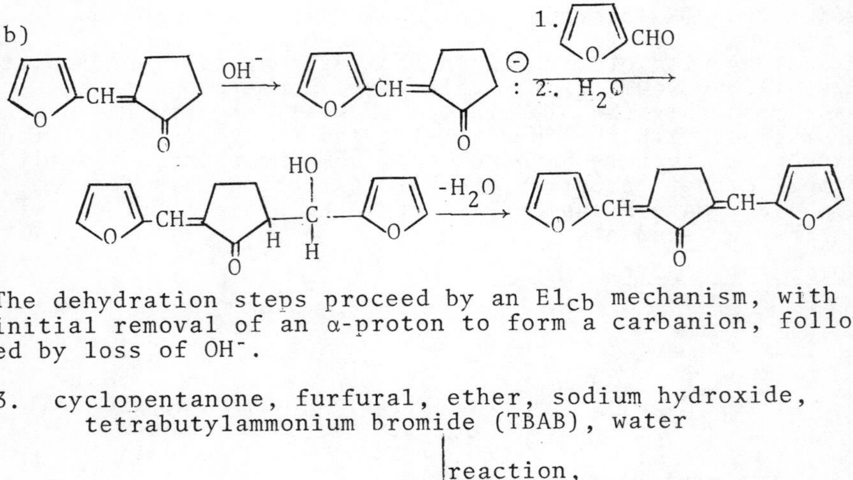

The dehydration steps proceed by an $E1_{cb}$ mechanism, with
initial removal of an α-proton to form a carbanion, followed by loss of OH^-.

3. cyclopentanone, furfural, ether, sodium hydroxide,
      tetrabutylammonium bromide (TBAB), water

reaction,
filter

filtrate   residue

furfurylidenecyclo-   difurfurylidenecyclo-
pentanone (FCP),   pentanone (DFCP), FCP
ether, NaOH, TBAB,
(cyclopentanone),
(furfural), (DFCP),   (Continued on next page)
water

74

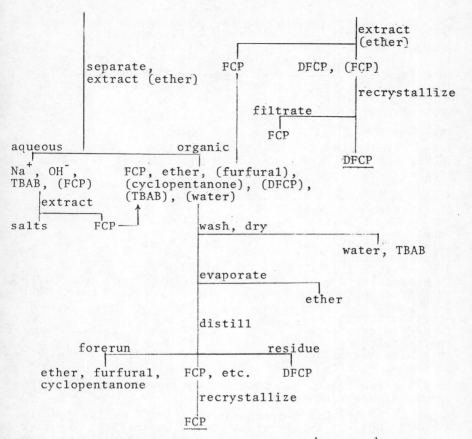

4. A possible mechanism is as follows ($Q^+ = Bu_4N^+$, Ar = furyl)

   1. $NaOH + Q^+Br^- \rightarrow Q^+OH^- + NaBr$ (aqueous phase)

   2.

   3.

Step 2 probably occurs in the aqueous phase or at the interface between the phases, from which Q^+ can carry the carbanion into the organic phase, where the condensation reaction (step 3) takes place. Q^+ then returns to the interface where step 2 is repeated, etc,

Diagram:

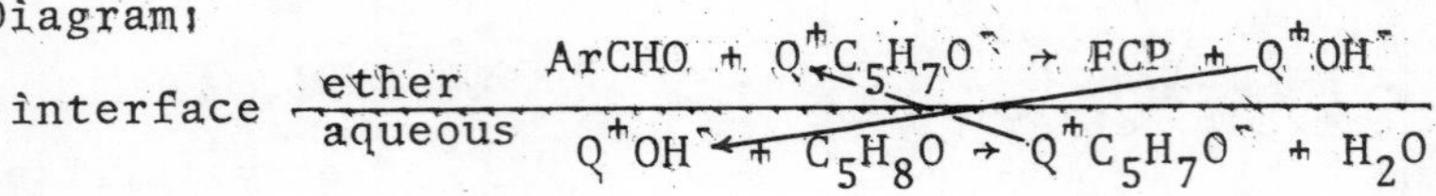

interface $\dfrac{\text{ether}}{\text{aqueous}}$ $\begin{array}{c} ArCHO + Q^+C_5H_7O^- \rightarrow FCP + Q^+OH^- \\ \hline Q^+OH^- + C_5H_8O \rightarrow Q^+C_5H_7O^- + H_2O \end{array}$

EXPERIMENT 36

Time: 1 period

Prelab calculations: 1.38 g p-nitroaniline; 1.44 g 2-naphthol; 8.3 ml 3M HCl; 20 ml 1M NaOH; 3.3 ml 3M sodium nitrite.

Chemicals and supplies per 10 students:

cloth, cotton or wool (1" x 2" strips)	0.1 sq yd
coupling components (from Table 1)	12-20 g each*
diazo components (from Table 1)	12-20 g each*
1M hydrochloric acid	240 ml (20 ml conc. HCl)
3M hydrochloric acid	340 ml (85 ml conc. HCl)
2-naphthol	20 g
p-nitroaniline	20 g
3M sodium carbonate	100 ml (32 g Na_2CO_3)
sodium chloride	50 g
1M sodium hydroxide	400 ml (16 g NaOH)
3M sodium nitrite	150 ml (31 g $NaNO_2$)

*at least two additional coupling components and two diazo components are recommended.

Comments: If the diazo component does not dissolve readily in the 3M HCl, it is all right to add up to 25 ml of water with heating and stirring. Some of the amine salt may precipitate on cooling but this should not affect the results. You may wish to provide most or all of the components in Table 1 and let students choose their own, so as to prepare as many different dyes as possible. Each student should ordinarily prepare Para Red and at least two additional dyes. (One of these should use an amine as the coupling component, so that the student has to apply the modification described in the Methodology.) Multifiber Fabric (E.V. 3) can be obtained from Testfabrics, Inc. P.O. Box 118, 200 Blackford Ave., Middlesex, NJ 08846.

Topics for Report:

1. Auxochromes like OH, NH_2, and NO_2 tend to deepen the color of the azo compound by increasing conjugation. Increasing the number of auxochromes or using more effective auxochromes thus changes the color of the dye in the direction yellow → orange → red. The effectiveness of various auxochromes is roughly in the order NO_2 > NMe_2 > NH_2 > OR (OH), with NO_2 having a particularly large effect when coupled

with an electron-donating group like OH. Para-substituted
auxochromes are much more effective than ortho or meta
substituted auxochromes. Extending the conjugation of the
chromophore (as in going from phenol to 2-naphthol as the
coupling component) also deepens the color of the dye.

2. General mechanism: $ArN{=}N: + $ ⬡$G \rightarrow ArN{=}N$⬡$\overset{+}{\underset{H}{}} G \xrightarrow{B:}$

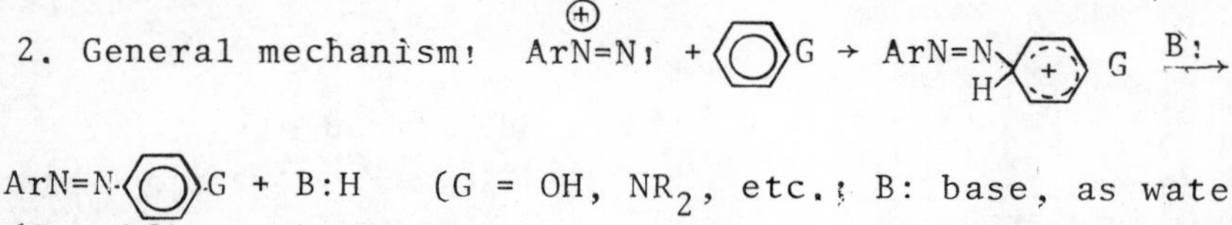

$ArN{=}N$⬡$G + B:H$ (G = OH, NR_2, etc.; B: base, as water)
(Specific mechanisms may differ in structures of aronium
ions)

3. Diazonium salts react with water at elevated tempera-
tures to yield phenols.

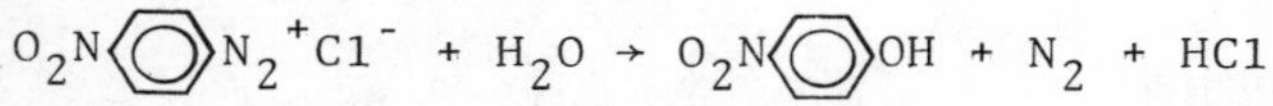

O_2N⬡$N_2^+Cl^- + H_2O \rightarrow O_2N$⬡$OH + N_2 + HCl$

4. Because the aryldiazonium ion is a comparatively bulky
nucleophile, substitution at the ortho position of N,N-
dimethylaniline is hindered by steric interference with the
amino substituent and para substitution is preferred. Substitution ortho to the oxygen of 2-naphthol is favored by
formation of the particularly stable quinonoid structure
shown below; substitution at the 4,5, or 7 position yields
aronium ions which cannot be represented by resonance
structures of this kind; substitution at the 3,6 and 8
positions yields aronium ions that do have such a resonance
structure, but in which the aromatic structure of the
second ring is destroyed. ArNN H

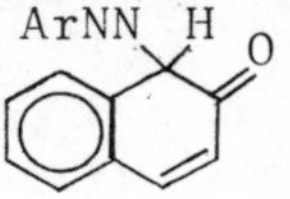

5. Aniline Yellow: $PhH \xrightarrow[H_2SO_4]{HNO_3} PhNO_2 \xrightarrow[HCl]{Sn} PhNH_2 \xrightarrow[NaNO_2]{HCl}$

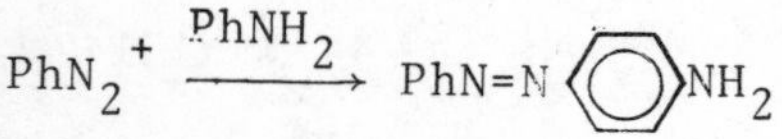

$PhN_2^+ \xrightarrow{PhNH_2} PhN{=}N$⬡$NH_2$

Bismark Brown: $PhNO_2 \xrightarrow[H_2SO_4]{HNO_3} O_2N$⬡$NO_2 \xrightarrow{Sn}{HCl} H_2N$⬡$NH_2 \xrightarrow[NaNO_2]{HCl}$
MPD

H_2N⬡$N_2^+ \xrightarrow{MPD} H_2N$⬡$\underset{H_3\overset{+}{N}}{N{=}N}$⬡$NH_2 \xrightarrow[NaNO_2]{HCl} \xrightarrow{MPD}$

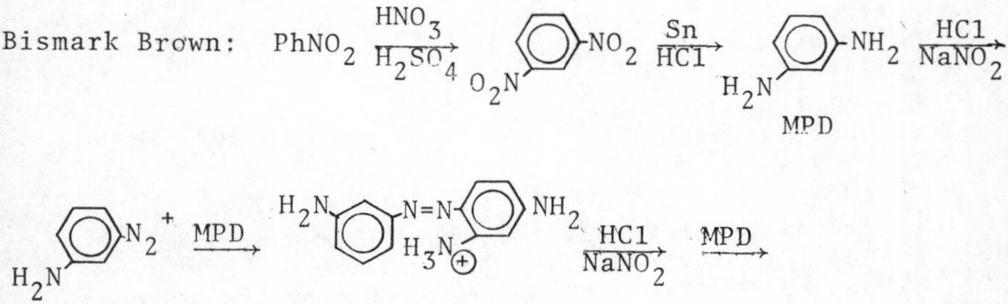

$$H_2N-\langle\bigcirc\rangle-N{=}N-\langle\bigcirc\rangle-N{=}N-\langle\bigcirc\rangle-NH_2$$

$\overset{+}{N}H_3 \qquad\qquad H_3\overset{+}{N}$

(MPD = meta-phenylenediamine)

MINILAB 15

Chemicals per 10 students:

ethanol, 95% or absolute	200 g (250 ml)
3M hydrochloric acid	120 ml (30 ml conc. HCl)
p-nitroaniline	20 g
potassium iodide	30 g
3M sodium nitrite	45 ml (9.3 g NaNO$_2$)

The melting point of p-iodonitrobenzene is 174°.

Reactions:

$$O_2N\langle\bigcirc\rangle NH_2 + 2HCl + NaNO_2 \rightarrow O_2N\langle\bigcirc\rangle N_2{}^+Cl^- + 2H_2O + NaCl$$

$$O_2N\langle\bigcirc\rangle N_2{}^+Cl^- + KI \rightarrow O_2N\langle\bigcirc\rangle I + N_2 + KCl$$

EXPERIMENT 37

Time: 2-3 periods

(Prelab Calculations should yield 3.38 g acetanilide and 1.88 g of the aminopyridine.)

Chemicals and Supplies per 10 students:

acetanilide (Eastman Kodak 3)	40 g
acetone, dried over K$_2$CO$_3$	315 g (400 ml)
85% acetone, aqueous	1.5 ℓ (1.0 kg acetone)
2-aminopyridine (Aldrich A7,799-7)	12 g
3-aminopyridine (Aldrich A7,820-9)	12 g
chlorosulfonic acid	220 g (125 ml)
Drierite, 10-20 mesh, no indicator	12 g
25% ethanol	800 ml (210 ml 95%)
70% ethanol	≈10 ml
6M hydrochloric acid	100 ml (50 ml conc. HCl)
magnesium sulfate, anhyd.	25 g
pyridine	30 g (30 ml)
dil. sodium hydroxide (gas traps)	1 ℓ (50 g NaOH)
3M sodium hydroxide	200 ml (24 g NaOH)
sulfanilamide	1 g
*culture plates, E. coli (etc.)	5
filter paper discs**	25
grease pencils	2
forceps	5

*Carolina Biological Supply Co., Burlington, NC 27215, cat. no. 15-5067 (or make your own agar plates)

**Cut with a #3 cork borer. These should be autoclaved or otherwise sterilized.

Special apparatus: incubating oven(s)

Comments: Part D can be omitted if desired, in which case the experiment can be completed in about 1½ periods. If part D is to be completed, it is important to use the purest available starting materials and to carefully purify the products; small amounts of impurities can drastically reduce their antibacterial activity. It may be advisable to recrystallize the Part B product from acetone and/or to carry out a second recrystallization of the final product if the initial results are not satisfactory.

It is important to get the ASC as dry as possible by blotting between filter papers, so that any residual water can be removed by the drying agents. Part B: when the ASC solution is mixed with 2-aminopyridine there is a short induction period after which the solution heats up suddenly and two layers separate; the oily bottom layer crystallizes after standing overnight, though traces of oil may remain. The reaction of ASC with 3-aminopyridine yields a solid product (rather than an oil) almost immediately. 3-sulfapyridine requires a relatively large volume of 85% acetone (usually 75 ml or more) for recrystallization - the usual mixed solvent technique of adding the solvents separately does not work. Because student yields of 2-sulfapyridine are rather low, it is suggested that 3-sulfapyridine be assigned to all but a few of your better students.

Topics for Report:

1. Activity against E. coli should be in the order 3-sulfanilamidopyridine > sulfanilamide > 2-sulfanilamidopyridine.

2. The free amino group would react with chlorosulfonic acid to yield PhNHSO$_3$H. Acetylating that group makes it less nucleophilic and moderates its activating effect on the benzene ring.

3. (a) ClSO$_3$H + H$_2$O → HCl + H$_2$SO$_4$; AcNH⟨◯⟩SO$_2$Cl + H$_2$O →

  AcNH⟨◯⟩SO$_2$OH + HCl

(b) H$_2$N⟨◯⟩SO$_2$NH⟨◯$_N$⟩ + OH$^-$ → H$_2$N⟨◯⟩SO$_2$N⟨◯$_N$⟩ + H$_2$O

H$_2$N⟨◯⟩SO$_2$NH⟨◯$_N$⟩ + H$^+$ → H$_2$N⟨◯⟩SONH⟨◯$_{+N-H}$⟩

(Pyridine nitrogen is more basic than ArNH$_2$)

4. A. ClSO$_3$H ⇌ HCl + SO$_3$; SO$_3$ + ArH → Ar$_{\overset{\oplus H}{SO_3^-}}$ $\xrightarrow{-H^+}$

ArSO$_3$H $\xrightarrow{\text{ClSO}_3\text{H}}$ ArSO$_2$Cl + H$_2$SO$_4$ (Ar = CH$_3$CONH⟨◯⟩-)

79

<u>B</u>. $ArSO_2Cl + Ar'NH_2 \rightarrow$ $Ar\overset{\underset{\displaystyle O}{\displaystyle O^{\ominus}}}{\underset{\displaystyle NH_2Ar'}{\overset{\displaystyle Cl}{S}}}$ $\xrightarrow{Py}$ $ArSO_2NHAr' + PyH^+Cl^-$

$\hspace{5cm} \oplus \hspace{1.5cm}$ (Ar' = 2- or 3-pyridyl)

<u>C</u>. $CH_3CONH\langle\bigcirc\rangle NHAr' + OH^- \rightarrow$ $CH_3\overset{\underset{\displaystyle OH}{}}{\overset{\displaystyle O^-}{C}}NH\langle\bigcirc\rangle NHAr' \rightarrow$

$CH_3\overset{O}{\overset{\|}{C}}-O^- + H_2N\langle\bigcirc\rangle NHAr'$ (after proton transfer)

The pentavalent species in B may be a transition state for S_N2 substitution rather than an intermediate.

5. $H_2N\langle\bigcirc\rangle SO_2NH_2 + Cl\langle\overset{\bigcirc}{N}\rangle \xrightarrow{B:} H_2N\langle\bigcirc\rangle SO_2NH\langle\overset{\bigcirc}{N}\rangle + HCl$

<u>Mechanism</u>: $H_2N\langle\bigcirc\rangle SO_2NH_2 \xrightarrow{B:} H_2N\langle\bigcirc\rangle SO_2NH^{\ominus} \xrightarrow{ArCl}$

$H_2N\langle\bigcirc\rangle SO_2NH\underset{Cl}{\overset{}{\langle\overset{\bigcirc}{N}\rangle}} \xrightarrow{-Cl^{\ominus}} H_2N\langle\bigcirc\rangle SO_2NH\langle\overset{\bigcirc}{N}\rangle$

6.   acetanilide, chlorosulfonic acid

trap   │reaction

HCl, SO₃   ASC, ClSO₃H (xs), acetanilide)
(as salts)

                           │water;
                           │filter

     filtrate   precipitate

ClSO₃H (as HCl,   ASC,(acetylsulfanilic acid),
H₂SO₄)   (water), (acetanilide)

                              │acetone,
                              │drying agents;
                              │filter

     residue   filtrate

acetylsulfanilic acid,   ASC, acetone, (acetanilide)
water

                                   │aminopyridine,
                                   │pyridine;
                                   │reaction

          APB*, acetone, pyridine, (ASC),
          (aminopyridine), (acetanilide)

*APB = 4-acetamido-N-
pyridylbenzenesulfonamide

(Continued on next page)

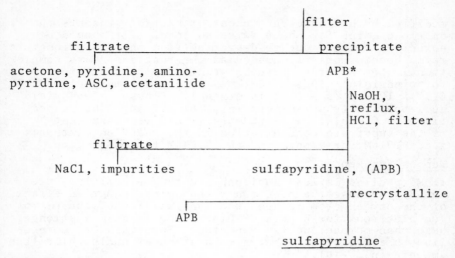

EXPERIMENT 38

Time: 1 period +

Chemicals and supplies per 10 students (working in pairs):

2-4-dinitrochlorobenzene	2 g
ethanol, absolute	260 g (330 ml)
ethanol, 95% (or abs.)	1.7 kg (2.2 ℓ)
morpholine	1.1 g (1.1 ml)
piperidine	0.4 g (0.5 ml)
sulfuric acid, conc.	45 g (25 ml)
volumetric flasks, 25 ml	5
volumetric pipets, 1-ml, 2-ml, 10-ml	5 ea.

Solution preparations: 2,4-dinitrochlorobenzene/ethanol - Accurately weigh 2.03 g of 2,4-dinitrochlorobenzene and make up to 100 ml in a volumetric flask with absolute ethanol. 1.0M morpholine/ethanol - Accurately weigh 2.18 g of morpholine and make up to 25 ml in a volumetric flask with absolute ethanol (for about 20 students). 0.40M piperidine/ethanol - Accurately weigh 0.85 g piperidine and make up to 25 ml in a volumetric flask with absolute ethanol (for about 20 students). 0.2M sulfuric acid/50% ethanol - Add 25 ml conc. sulfuric acid to 1.2 liters of 95% ethanol and make up to 2.3 liters with distilled water.

Comments: Each pair of students will need at least 8 125-ml Erlenmeyer flasks and 7 15-cm test tubes or the equivalent. If thermostat baths are available, the reactions should be run at 25⁰. If time permits, it is best to let the reaction mixtures stand at room temperature for 48 hours or longer before getting the A_∞ readings; the morpholine reaction is only about 95% complete after 2 hours at 50⁰, and the piperidine reaction 98%. The experiment can be

modified by having each student or pair of students run
only one amine for about twice as long. (It should be
noted that errors due to deviation from the stoichiometric
ratio become more pronounced as the reaction nears completion.) Second-order rate constants from the literature are,
for piperidine: 19.5×10^3 $M^{-1}s^{-1}$, and for morpholine,
4.17×10^{-3} $M^{-1}s^{-1}$ (25^0 in ethanol). These values yield
nucleophilicity parameters, n_{Ar}, of 3.7 and 3.0, respectively. Calculations: Students should remember that S_0
is the substrate concentration in the reaction mixture;
i.e. 0.016M for piperidine and 0.040M for morpholine.

Topics for Report:

1. Morpholine is less nucleophilic than piperidine, presumably because the electron-withdrawing inductive effect
of the oxygen atom reduces the availability of the nitrogen electrons for bonding. Piperidine is also a stronger
base than morpholine, as expected. Surprisingly, morpholine is a weaker base but a much stronger nucleophile than
ammonia (n_{Ar} = 0).

2. $O_2N\langle\bigcirc\rangle Cl$ + $HN\bigcirc$ $\rightarrow$ $\longrightarrow$

NO_2 + Cl^- + $\langle\ \rangle NH_2^+$ (and an analogous mechanism for
morpholine)

3. (a) A = εbc (Beer's law). At any time t, the molar concentration (c) of the product in the cuvette is 0.0035x
(see Topic 5); when the reaction is complete, it is
$0.0035S_0$. Thus A = $0.0035\varepsilon bx$; $A_\infty = 0.0035\varepsilon bS_0$; and A_∞ - A =
$0.0035\varepsilon b(S_0 - x)$, which rearranges to give Equation 5 with
q = $0.0035\varepsilon b$.   (b) If $N_0 = 2S_0$, Equation 3 becomes dx/dt =
$k(S_0 - x)(2S_0 - 2x) = 2k(S_0 - x)^2$. Rearranging gives
$dx/(S_0 - x)^2$ = 2kdt and integrating this equation yields
Equation 4.

4. ArCl + $HOCH_2CH_2OH$ $\xrightarrow{Na}$ $ArOCH_2CH_2OH$ $\xrightarrow{Na}$
(Ar = picryl)

5. (a) After dilution, c = $S_0 \times 1/26 \times 1/11 = 0.0035S_0$,
which is 5.59×10^{-5} for piperidine and 1.40×10^{-4}M for
morpholine.

(b) The ½" "test-tube" sample cells ordinarily used in the
Spectronic 20 have an internal diameter of 1.16 cm (square
cuvettes with the same I.D. are also available). Absorp-
tivity = A_∞/bc; a reported value for 2,4-dinitrophenyl-
piperidine is 15,400.

6. $Y\langle\ \rangle NH$ + H_2SO_4 $\rightarrow$ $Y\langle\ \rangle NH_2^+$ + HSO_4^-. The amine salt has no
unshared electrons with which to attck the DNCB; i.e. it is
no longer nucleophilic.

MINILAB 16

Chemicals per 10 students:

2,4-dinitrochlorobenzene	7 g
ethanol, 95%	100 g (130 ml)
ethyl acetate	270 g (300 ml)
hydrazine hydrate (64% hydrazine)	7 g (7 ml)

Mechanism:

$$O_2N\text{—}C_6H_3(NO_2)\text{—}Cl + NH_2NH_2 \rightarrow O_2N\text{—}C_6H_3(NO_2)\text{—}\overset{Cl}{\underset{\overset{+}{NO_2}}{NH_2NH_2}} \xrightarrow{B:}$$

$$O_2N\text{—}C_6H_3(NO_2)\text{—}NHNH_2 + BH^+ + Cl^-$$

(B: = water or ethanol)

EXPERIMENT 39

Time: 2-3 periods

Prelab calculations: 9.82 g (10.4 ml) cyclohexanone; 10.45 g (10.5 ml) morpholine; 12.14 g (16.7 ml) triethyl-amine; 14.32 g (14.2 ml) propionic anhydride.

Chemicals per 10 students:

cyclohexanone	125 g (130 ml)
methylene chloride	3 kg (2.25 ℓ)
morpholine	150 g (150 ml)
propionic anhydride	185 g (185 ml)
toluene	*730 g (840 ml)
p-toluenesulfonic acid	1.5 g

*assumes about 45 ml for each water separator

Comments: The long reflux period in Part A should be per-formed, if possible, during another experiment or in con-junction with a minilab.* For example, the kinetic runs of Exp. 38 could be carried out after the reflux is begun, and absorbance values measured during one of the Part B re-fluxes. No special drying should be necessary if freshly opened, reagent grade solvents and reactants are used.
E.V. 1: The IR and NMR spectra of the similar 2-acetylcyclo-hexanone are reproduced in the Aldrich collections: IR (2nd ed) p. 227C, NMR 2, p. 123D, and also in the Sadtler Spectra: IR 36702, IR (grating) 14691, NMR 7941. The NMR spectrum in carbon tetrachloride shows nearly 100% enol, with the enol OH signal at 15.8δ.

Topics for Report:

1. Water hydrolyzes the enamine back to the starting materials:

$$\text{(cyclohexenyl)—N}\bigcirc\text{O} + H_2O \rightarrow \text{(cyclohexanone)}{=}O + O\bigcirc NH$$

2. (a)

$$\text{enamine}^{\oplus}{:}{\ominus} + (CH_3CH_2CO)_2O \rightarrow N^{\oplus}\text{—}\overset{O\ominus}{\underset{CH_2CH_3}{C\text{—}OCOCH_2CH_3}} \rightarrow$$

enamine resonance structure

*The O.S. procedure for 1-morpholinecyclohexene calls for a 3-hour reflux, but we have found the reaction to be nearly complete after 1 hour.

83

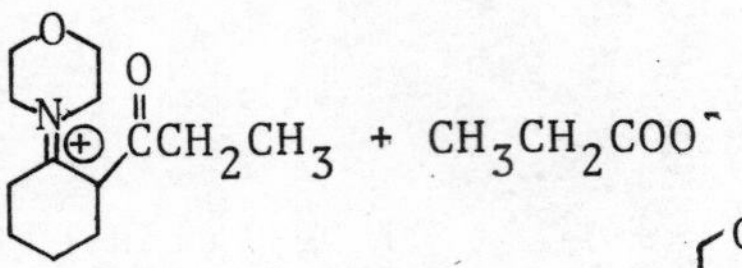

(b) Product (a) + H_2O →

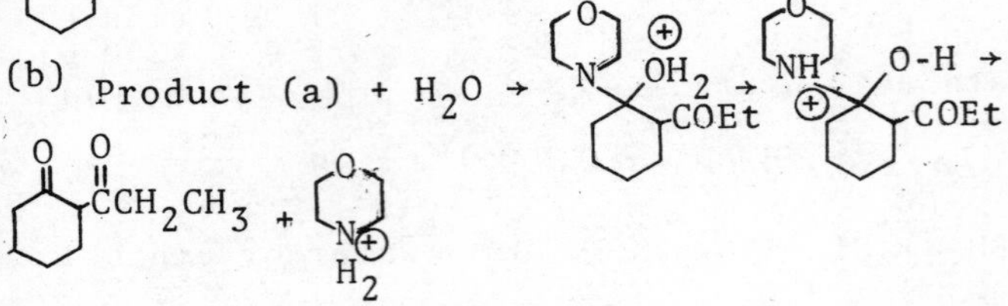

3. (a)

$$\text{A} \qquad \text{B} \qquad \text{C} \qquad \text{D}$$

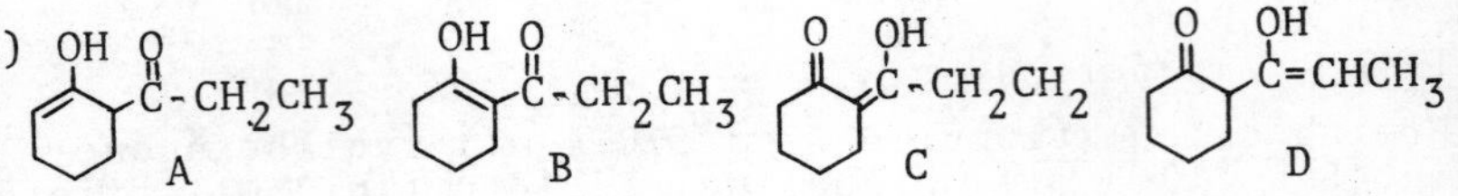

(b) B, because the carbon-carbon double bond is endocyclic and conjugated with a carbonyl group. A and D are less stable because they are not conjugated, C because the double bond is exocyclic.

4.   cyclohexanone, morpholine, toluene, TsOH

|
reflux

enamine, toluene, morpholine (xs), TsOH, (cyclohexanone)

|
distill

forerun ——————————|—————————————— residue

morpholine, toluene   enamine   TsOH
cyclohexanone

|
prop. anhydride, etc.;
reflux

iminium slat, propionate ion, triethylamine,
methylene chloride, propionic anhydride (xs), (enamine)

|
evaporate

methylene chloride ——————|

|
water;
reflux

2-propionylcyclohexanone, morpholine, propionate,
triethylamine, water, (cyclohexanone)

|
extract (CH_2Cl_2)

(Continued on next page)

84

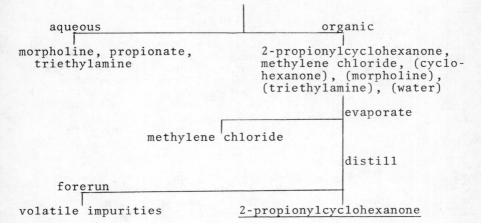

```
              aqueous                          organic
    morpholine, propionate,          2-propionylcyclohexanone,
       triethylamine                 methylene chloride, (cyclo-
                                     hexanone), (morpholine),
                                     (triethylamine), (water)

                                                evaporate

                    methylene chloride

                                                distill

         forerun

    volatile impurities       2-propionylcyclohexanone
```

EXPERIMENT 40

Time: 1 period

Prelab calculations: 6.61 g (5.7 ml) dimethyl malonate;
5.11 g (6.0 ml) mesityl oxide

Chemicals and supplies per 10 students:

acetone	140 g (175 ml)
calcium chloride or Drierite	450 g
chloroform-d	10 g (7 ml)
dimethyl malonate	85 g (75 ml)
(Aldrich 13,644-1)	
6M hydrochloric acid	400 ml (200 ml conc. HCl)
mesityl oxide	65 g (75 ml)
(Aldrich M785-5)	
methanol	30 g (40 ml)
3M sodium hydroxide	500 ml (60 g NaOH)
25% sodium methoxide/methanol	150 g (160 ml)
(Aldrich 15,625-6)	
tetramethylsilane	0.15 g (0.2 ml)
NMR tubes	10

Special apparatus: NMR spectrometer(s)

Comments: The sodium methoxide/methanol solution can be
prepared by adding sodium to methanol, but it is much
simpler (and safer) to buy it ready-made. A solid forms
while the mesityl oxide is added, but it should dissolve
when heating is begun. The ester intermediate (as the
sodium salt) precipitates as a white solid during the
initial reflux period. Purification: Dimedone can form
supersaturated solutions, so the recrystallization flask
should be scratched during the cooling period. The filtrate can be concentrated for a second crop of crystals -
these are generally a darker yellow than the first crop.
The total yield should be about 5 g of nearly white

crystals. <u>Analysis</u>: If NMR spectrometers are not available for student use, the spectrum in the back of this manual can be photocopied and distributed (see Topic 1 for interpretation). <u>E.V. 1</u>: The IR spectrum is reproduced in the Aldrich collection (2nd ed.) p. 227B. The strong enolic O-H band centered near 2500 cm^{-1} and the lack of a diketone carbonyl band confirms that the enol form predominates. (Note that the carbonyl band of an enolic ketone occurs at 1540-1640 cm^{-1}.) <u>E.V. 2</u>: The NMR spectrum in a mixture of DMSO-d$_6$ and chloroform-d is reproduced in the Aldrich NMR collection (<u>2</u>, 123C): it shows almost 100% enol form.

Topics for Report:

1. The enol OH signal should be at about 10.9δ, the enol H-C=C signal at 5.5δ, and the keto H-C(C=O)$_2$ at 3.4δ. (In addition, the methylene protons of the enol form absorb at 2.3δ and the corresponding protons of the keto form at 2.6δ; their area ratio should <u>also</u> equal K.)

2. (a) $CH_2(COOMe)_2 \xrightarrow{\text{NaOMe}} {}^{\ominus}CH(COOMe)_2$;

$CH_3\overset{O}{\overset{\|}{C}}CH=\overset{|}{\underset{CH_3}{C}}CH_3 + {}^{\ominus}CH(COOMe)_2 \rightarrow CH_3\overset{O}{\overset{\|}{C}}{=}CH\overset{CH_3}{\underset{\ominus}{\overset{|}{C}}}-CH(COOMe)_2 \xrightarrow{\text{MeOH}}$

$CH_3\overset{O}{\overset{\|}{C}}CH_2\overset{CH_3}{\underset{CH_3}{\overset{|}{C}}}-CH(COOMe)_2$ (via protonation of O and rearrangement of enol form)

<u>A</u>

(b) <u>A</u> $\xrightarrow{\text{NaOMe}}$ [structure: $O=C$ with $CH_2^{\ominus}$, C-OMe, H_2C, CH-COOMe, ring with C and Me Me] $\rightarrow$ [structure: $O=C$ with CH_2, $\overset{O^{\ominus}}{C}$-OMe, H_2C, CH-COOMe, ring with C and Me Me] $\rightarrow$

[structure: $O=C$, CH_2, $C=O$, H_2C, CH-COOMe, ring with C and Me Me] $+ MeO^-$

<u>B</u>

(c) + CO$_2$ (followed by enol-keto rearrangements)

<u>C</u>

3. Mesityl oxide = 4-methyl-3-penten-2-one; dimethyl malonate = dimethyl 1,3-propanedioate. The intermediates and products are, in order of formation, dimethyl 2-(1,1-dimethyl-3-oxobutyl)-1,3-propanedioate; methyl 6,6-dimethyl-

2,4-dioxocyclohexanecarboxylate; 6,6-dimethyl-2,4-dioxo-cyclohexanecarboxylic acid; and 5,5-dimethyl-1,3-cyclo-hexanedione.

4.   dimethyl malonate, mesityl oxide, MeOH, NaOMe

|reflux

      *B (salt), MeOH, (NaOMe), (mes. oxide),
       (dimethyl malonate), (by-products)

*see Topic 2   _____|evaporate
for structures |
              MeOH

                      |NaOH:
                      |reflux

      *C (salt), NaOH, (NaOMe),(MeOH), (mes. oxide),
            (dimethyl malonate), (by-products)

            |_____|HCl, boil
CO_2, MeOH

            dimedone, Na^+, Cl^-, (mes. oxide),
               (dimethyl malonate), (by-products)

                      |filter

      filtrate   precipitate
      |_____|
Na^+, Cl^-, mes. oxide,   dimedone, (by-products)
dimethyl malonate
                                             |recrystallize
            |_____|
            by-products
                                             |
                                          dimedone

(by-products may include unreacted A, B or C as well as
substances arising from alternate reaction paths)

5.
   $CH_3CH_2CH_2CH(-\bigcirc\hspace{-1.5em}\times)_2$ or equivalent enolic forms.

MINILAB 17

Chemicals per 10 students:

aniline	7 g (7 ml)
chalcone (Aldrich 13,612-3)	15 g
absolute ethanol	200 g (260 ml)

It is important that the reactants be pure; the aniline, in particular, should be recently distilled.

Mechanism:

$$Ph\overset{\overset{\text{O}}{\|}}{C}-CH=CHPh + PhNH_2 \rightarrow Ph\overset{\overset{:\overset{\ominus}{O}}{\|}}{C}=CHCH-\overset{\oplus}{N}H_2Ph \rightarrow$$
$$\underset{Ph}{|}$$

$$Ph\overset{\overset{\text{OH}}{|}}{C}=CHCH-NHPh \rightarrow Ph\overset{\overset{\text{O}}{\|}}{C}-CH_2CH-NHPh \quad \text{(1,4-addition followed by}$$
$$\underset{Ph}{|} \qquad\qquad \underset{Ph}{|} \qquad \text{enol-keto rearrangement)}$$

EXPERIMENT 41

Time: 1 period

Prelab calculations: 3.56 g anthracene; 7.00 g chromium trioxide

Chemicals per 10 students:

acetic acid, glacial	775 g (715 ml)
anthracene (Aldrich A8,920-0)	45 g
chromium trioxide (Baker 1638)	90 g
1M sodium hydroxide	450 ml (18 g NaOH)

Comments: The anthracene need not dissolve completely before the chromium trioxide addition is begun. Separation: It is recommended that chemically resistant filter paper (e.g., Whatman No. 111) be used in the vacuum filtration, since ordinary filter paper may tear during the washing steps. The filtration is also quite slow with Whatman No. 1 paper. The acetic acid in the filtrate should be removed before the washings are begun. Purification: Without suitable apparatus, the sublimation is quite tedious and the yield low, so students may purify only a half gram or so for the melting point. E.V. 1: An alternative purification procedure, suggested by David Todd of Worcester Polytechnic Institute, is to bring the product to a gentle boil in enough dioxane (hot plate or other electric heater required) to dissolve it, and let it stand overnight to crystallize. The filtrate can be concentrated for a second crop, if desired. E.V. 3: The UV spectra have the following absorption maxima. Anthraquinone: 252 (4.7), 278 (4.1), 330 (3.7); anthracene: 310 (3.11), 324 (3.45), 340 (3.73), 357 (3.89), 376 (3.87). (Numbers in parentheses are log ε values.)

Topics for Report:

1. The stoichiometric quantities of CrO_3 and acetic acid are 40 mmol and 120 mmol, respectively (2 moles CrO_3 and 6 moles HOAc per mole anthracene).

CrO_3: 70 mmol used - 75% excess. HOAc: 55 ml (57.7 g, 960 mmol) - 700% excess.

2.

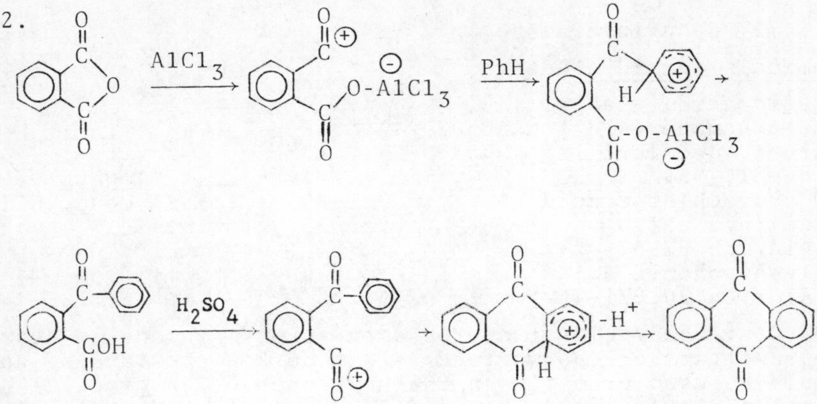

Friedel-Crafts reaction. (acylium ions are probably complexed with the aluminum salts)

3.   anthracene, acetic acid, CrO_3

|reflux

anthraquinone, acetic acid, CrO_3 (xs), Cr^{3+}
            OAc^-, water, anthracene

|filter;
|wash (NaOH, H_2O)

filtrate   precipitate

acetic acid, CrO_3,   anthraquinone, (water),
Cr^{3+}, OAc^- (anthracene)

|dry, sublime

anthraquinone

4. $2H_2SO_4 \rightleftharpoons SO_3 + H_3O^+ + HSO_4^-$;

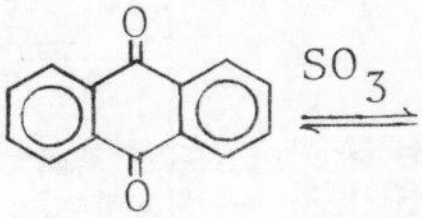

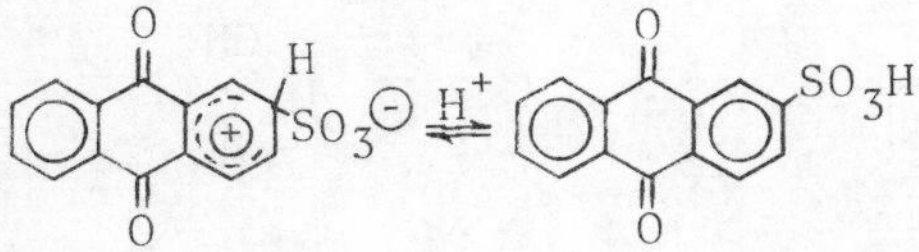

EXPERIMENT 42

Time: 1-1½ periods

Prelab calculations: 3.01 g (2.9 ml) acetophenone; 2.70 g

89

(2.5 ml) pehnylhydrazone

Chemicals per 10 students:

acetic acid, glacial	4 g (4 ml)
acetophenone (Aldrich A1,070-1)	40 g (40 ml)
carbon, decolorizing	3 g
ethanol, 95%	1 kg (1.4 ℓ)
1M hydrochloric acid	180 ml (15 ml conc. HCl)
phenylhydrazine	35 g (32 ml)
(Aldrich P2,625-2)	
polyphosphoric acid	700 g (330 ml)
(Aldrich 20,821-3)	

Comments: The yield of acetophenone phenyldrazone is near-ly quantitative; the white crystals darken on standing and should be used promptly in the next step. The crude 2-phenylindole is greenish and filters slowly. Purification: The recrystallization is challenging, in that phenylindole dissolves very slowly in boiling ethanol and it is nearly impossible to get it all in solution. Probably the best method is to boil it in 75-100 ml of ethanol for 15-20 minutes, filter the hot solution (washing the residue with hot ethanol), and saturate the filtrate with hot water. Adding carbon does not appear to remove much color from the filtrate, but it does give a better-looking product; phenylindole should crystallize as shining white plates from the green solution. If the undissolved residue and mother liquor are worked up, the yield should be ≃75%. E.V. 1: The IR spectrum is reproduced in the Aldrich col-lection (2nd ed.) p. 1080F. E.V. 2: Most indoles give a red color with this test.

Topics for Report:
1.

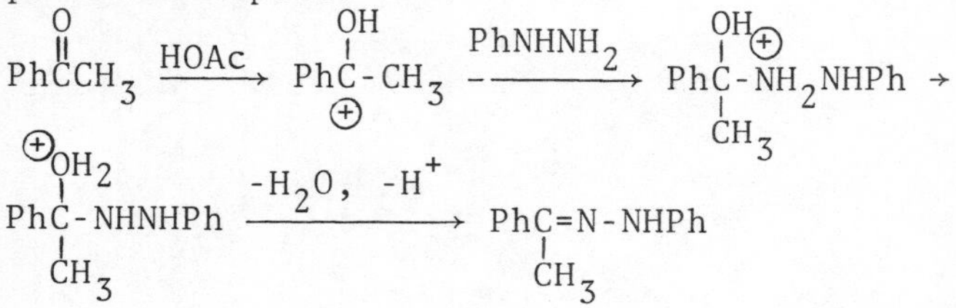

2. This is the Robinson mechanism (illustrated in the text) with R = H and R' = phenyl.

Transition states: (3)

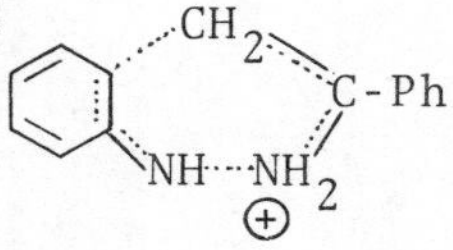

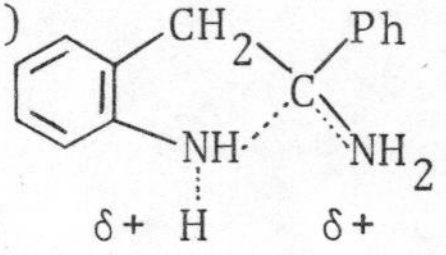

3.   acetophenone, phenylhydrazine,
               ethanol, (acetic acid)

(Continued on next page)

90

```
                              │heat
acetophenone phenylhydrazone,│ethanol, (acetic acid),
    (acetophenone), (phenylhydrazone), (water)

                              │filter, wash (HCl, EtOH)
  filtrate                    │        precipitate
┌─────────────────────────────┴──────────────────────┐
ethanol, water, acetic acid,        acetophenone phenyl-
acetophenone, phenylhydrazone            hydrazone

                                    │PPA, heat
                  2-phenylindole, PPA, $NH_4^+$,
                        (phenylhydrazone)

                                    │water; filter
       filtrate                     │
┌──────────────────────────────────┴──────────┐
PPA, $NH_4^+$                                  │recrystallize
                                    ┌──────────┴─────────┐
                  2-phenylindole              phenyl-
                                              hydrazone
```

4.

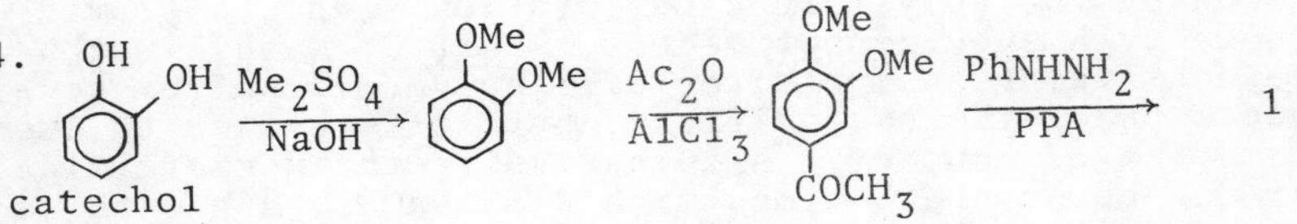

catechol

Me$_2$SO$_4$ / NaOH → Ac$_2$O / AlCl$_3$ → PhNHNH$_2$ / PPA → 1

MINILAB 18

Chemicals per 10 students:

acetone	20 g (25 ml)
ethanol, 95%	20 g (25 ml)
ethyl ether	18 g (25 ml)
o-nitrobenzaldehyde	7 g
(Aldrich N1,080-2)	
1M sodium hydroxide	25 ml (1 g NaOH)

EXPERIMENT 43

Time: 1 period

Chemicals and supplies per 10 students:

benzoyl peroxide	2 g
2-butanone	20 g (15 ml)
methanol	600 g (750 ml)
poly(methyl methacrylate)	125 g
(Aldrich 18,223-0)	
10% potassium aluminum	150 ml (15 g alum)
sulfate, aq. soln.	

91

```
potassium peroxydisulfate          1 g
sodium lauryl sulfate              3 g
styrene (Aldrich S497-2)         145 g (160 ml)
toluene                           55 g (65 ml)

magnetic stirrers                 10
magnetic stirring bars            10
microscope slides                 10
razor blades                       2
```

Special apparatus: infrared spectrophotometer(s); nitrogen gas bottles

Comments: Part A: The magnetic stirrer can be omitted if the mixture is shaken frequently enough to maintain an emulsion. Other detergents, or soaps such as sodium stearate, can be used in place of sodium lauryl sulfate. Inhibitor should be removed by distilling the styrene or washing it with 10% NaOH shortly before the laboratory period. (The inhibited monomer will polymerize but the yield after 3 hours is low.) Part B: The benzoyl peroxide should be put out in small quantities in containers like plastic-capped vials or paper cups, and students should pour it directly from these containers onto glassine papers. If a 50 mg sample is provided for comparison, students can estimate the proper quantity without weighing it out. The styrene should be distilled or washed (10% NaOH) and dried before use. The infrared spectrum can be compared with that given by the polystyrene calibration standard provided with most IR spectrophotometers. Part C: Methyl methacrylate containing an inhibitor will polymerize after several weeks under these conditions, but it is better to use freshly prepared monomer. Alternatively, the commercial monomer can be washed 3 times with 0.5% NaOH, then with water, and dried over magnesium sulfate.

Topics for Report:

1. $(PhCOO)_2 \xrightarrow{heat} 2PhCOO \cdot \rightarrow 2Ph \cdot \ 2CO_2$

$Ph \cdot + CH_2 = \underset{\underset{CH_3}{|}}{C} - COOMe \rightarrow PhCH_2 \underset{\underset{CH_3}{|}}{\overset{\cdot}{C}} - COOMe$

$PhCH_2 \underset{\underset{CH_3}{|}}{\overset{\cdot}{C}} - COOMe + CH_2 = \underset{\underset{CH_3}{|}}{C} - COOMe \rightarrow PhCH_2 \underset{\underset{CH_3}{|}}{C} - CH_2 \underset{\underset{CH_3}{|}}{\overset{\overset{COOMe}{|}}{\overset{\cdot}{C}}} - COOMe$, etc. (plus

chain-terminating steps)

2. 1. $CHCl = CHCl$ 2. $CF_2 = CFCl$ 3. $\underset{CH_2 - CH_2}{NH}$ 4. $CH_2 = \underset{\underset{CH_3}{|}}{C} - CH = CH_2$

   5. $CH_2 = CHCH_2CH_3$ 6. $CH_2 = CHCN$ and $CH_2 = CHPh$

3.

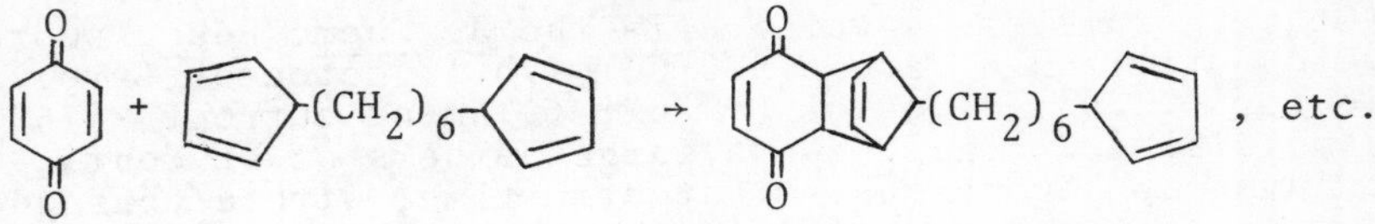

Repeating unit:

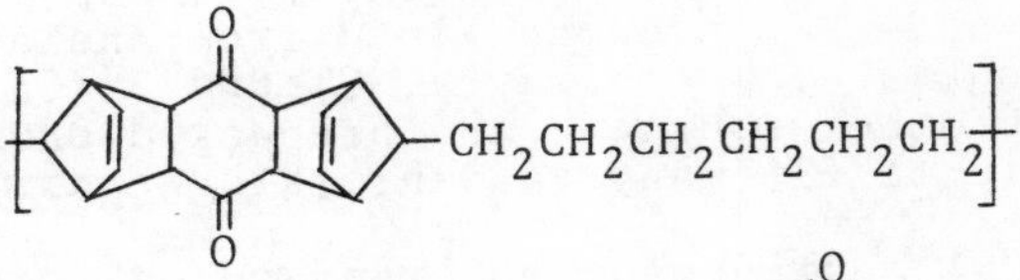

4. $nHOCH_2CH_2OH \rightarrow (-CH_2CH_2O-)_n + nH_2O$; $nCH_2-CH_2 \rightarrow$

       condensation   addition

$(-CH_2CH_2O-)_n$

MINILAB 19

Chemicals per 10 students:

castor oil (Baker 1518)	70 g (75 ml)
glycerol	15 g (12 ml)
silicone oil, DC-200 (Baker U964)	1½ g (1½ ml)
tolylene-2,4-diisocyanate (Aldrich 21,683-6)	50 ml (60 g)
triethylamine	1 g (1½ ml)

EXPERIMENT 44

Time: 1 period

Chemicals and supplies per 10 students:

boron trifluoride-methanol complex (Aldrich 13,482-1)	16 g (13 ml)
chloroform	1 g (0.7 ml)
cooking oil	2 g (2 ml)
methanol	100 g (130 ml)
methyl heptadecanoate (Fluka 51640)	0.6 g (0.7 ml)
petroleum ether (30-60°)	275 g (415 ml)
sodium chloride, saturated soln.	250 ml (≈80 g NaCl)

Solution preparations: 12.5% boron trifluoride/methanol - Mix 13 ml of commercial boron trifluoride-methanol complex with 65 ml of dry methanol. 0.5M sodium hydroxide/methanol Dissolve 1.3 g NaOH pellets in methanol, make up to 65 ml with additional methanol.

Special apparatus: gas chromatograph(s) with DEGS columns

Comments: The experiment is more interesting if students are allowed to bring their own brands of cooking oil (or oleomargarine - see E.V. 1) for testing. Analysis: The column should be 6-10' long, packed with 10% DEGS on 60-80 mesh Chromosorb W, and the analysis should be carried out at 185-90° at a helium flow rate of about 120 ml/min.

(Other parameters specified in the J. Chem. Educ. source, which utilized a Varian 920 GC with TC detector, are: injector - $228°$; detector - $200°$; filament current - 150 mA; sample size 2-3 μl.) With large numbers of students, the column may become contaminated in time, with a consequent decrease in separation efficiency. Stabilized DEGS stationary phases can be obtained from Analabs, Inc., 80 Republic Drive, North Haven, CT 06473 (Cat. No. SLP-027) and many other sources, and prepacked DEGS columns are available to fit many gas chromatographs.

The methyl heptadecanoate standard can, if desired, be made by esterifying heptadecanoic acid in BF_3/methanol. Using peak areas for quantitative analysis gives reasonably good results, but for more precise values the peak areas may be multiplied by the following detector response factors: myristic - 0.908; palmitic - 0.954; stearic - 1.010; oleic 0.980; linoleic - 1.071; linolenic - 1.172 (these apply only to TC detectors). The area of the methyl heptadecanoate peak should not be counted in computing weight percentages. Approximate compositions of different classes of fats and oils are given in the J. Chem. Educ. source (51, 406 (1974)).

Topics for Report:

1. PUFA/SFA = (peak areas for linoleic & linolenic esters)/ (peak areas for palmitic, stearic, lauric, and myristic esters). The latter two saturated esters may not be present in detectable quantities. PUFA/SFA ratios for a number of commercial cooking oils and fats are given in Consumer Reports, Sept. 1973, pp. 556-7 and can be used to check student results. Presumably the oils with the highest PUFA/ SFA ratios should confer the greatest health benefit, though this is still a matter of controversy.

2. lauric acid - $CH_3(CH_2)_{10}COOH$; dodecanoic acid; 12:0

myristic acid - $CH_3(CH_2)_{12}COOH$; tetradecanoic acid; 14:0

palmitic acid - $CH_3(CH_2)_{14}COOH$; hexadecanoic acid; 16:0

palmitoleic acid - $CH_3(CH_2)_5CH=CH(CH_2)_7COOH$; 9-hexadecenoic acid; 9-16:1

stearic acid - $CH_3(CH_2)_{16}COOH$; octadecanoic acid; 18:0

oleic acid - $CH_3(CH_2)_7CH=CH(CH_2)_7COOH$; 9-octadecenoic acid; 9-18:1

linoleic acid - $CH_3(CH_2)_3(CH_2CH=CH)_2(CH_2)_7COOH$; 9,12-octadecadienoic acid; 9,12-18:2

linolenic acid - $CH_3(CH_2CH=CH)_3(CH_2)_7COOH$; 9,12,15-octadecatrienoic acid; 9,12,15-18:3

(all double bonds are cis)

3. A typical vegetable oil contains mostly C-18 fatty acid glycerides; corn oil is about 2% stearic acid, 25% lino-

leic acid, and 62% linolenic acid; soybean oil about 4% stearic, 43% linoleic, and 40% linolenic. Hydrogenation of both linoleic and linolenic acid yields stearic acid.

4. Glycerol is water soluble and petroleum ether insoluble, so it remains in the aqueous layer during the extraction.

5.
$$CH_3(CH_2)_{10}COOCH_2$$
$$CH_3(CH_2)_{10}COOCH \xrightarrow[\underline{or\ LiAlH_4}]{Na,\ EtOH} CH_3(CH_2)_{10}CH_2OH \xrightarrow{H_2SO_4}$$
$$CH_3(CH_2)_{10}COOCH_2 \qquad\qquad + glycerol$$

$$CH_3(CH_2)_{10}CH_2OSO_3H \xrightarrow{NaOH} CH_3(CH_2)_{10}CH_2OSO_3^-Na^+$$

PART III

SYSTEMATIC ORGANIC QUALITATIVE ANALYSIS

Students should be provided with 8-10 ml of a liquid unknown or 4-5 g of a solid, and should ordinarily be allowed to obtain more if they run out. A larger quantity of
an ester or amide that requires hydrolysis may be needed.
If time permits, several unknowns from specified families
(e.g. alcohols, amines, and carboxylic acids) can be issued,
followed by 2-3 general unknowns from the families specified in the Methodology. Suitable unknowns are listed in
the Tables in Appendix VIII. The time required for iden-
tification can vary considerably, but it is best to allow
about 1 period for a specific unknown and 1½-2 periods for
each general unknown. Although the use of spectrometric
methods (especially IR) is recommended, the unknowns can
be identified without any instruments. It is a good
practice to have students report their observed melting
point and boiling point values so that they can purify
their unknowns (or repeat the determinations) if their results are too far from the literature values.

The following quantities of chemicals and supplies are for
10 tests or derivative preparations; if controls are used,
a student may repeat each classification test several
times, so the quantities must be adjusted accordingly.
For example, a student with an aromatic amine who runs
the quinhydrone test should compare the colors given by
aniline, N-methylaniline, and N,N-dimethylaniline with
that produced by his unknown, so 10 students may consume
roughly four times the quantities listed. (Controls are
necessary for this test; they are not essential for many
tests in which the occurrence of a precipitate or color
change clearly distinguishes a positive test from a negative one.) In derivative preparations D-1, D-2, D-9, D-10,
D-14, D-20, and D-24 either or both of the indicated deriv-
atizing reagents can be provided. Because of wide variations in the amounts of chemicals (particularly recrystallization solvents) needed for many of the procedures, the
quantities given should be regarded as rough approxima-
tions.

An asterisk designates a solution or reagent for which a preparation is described following the chemical lists. All other solutions are aqueous, with the amount of solute given in parentheses following the volume of the solution. All chemical catalog numbers are from Aldrich Chemical Co. unless otherwise indicated.

Chemicals and supplies per 10 tests

Solubility Tests

5% hydrochloric acid	40 ml (4.7 ml conc. HCl)
5% sodium bicarbonate	30 ml (1.55 g $NaHCO_3$)
5% sodium hydroxide	30 ml (1.58 g NaOH)
sulfuric acid, conc.	20 ml (37 g)

Classification Tests

C-1 acetyl chloride (11,418-9)   10 ml
     ammonia, conc. aq.   7.5 ml
     N,N-dimethylaniline   4 ml
     control: 1-butanol   7.5 ml

C-2 10% copper(II) sulfate   1.5 ml (0.26 g $CuSO_4 \cdot 5H_2O$)

 (a) 6M hydrochloric acid   60 ml (30 ml conc. HCl)
     *20% potassium hydroxide/glycerine 20 ml
     6M sodium hydroxide   60 ml (14.4 g NaOH)
     controls: acetanilide, benzamide   1.5 g ea

 (b) potassium carbonate   100 g
     6M sodium hydroxide   150 ml (36 g NaOH)
     6M sulfuric acid   75 ml (25 ml conc. H_2SO_4)
     controls: ethyl benzoate,   15 ml ea
       butyl acetate

C-3 aluminum chloride, anhyd.   3 g
     chloroform   30 ml
     controls: biphenyl, toluene   1.5 g/2 ml ea

C-4 *acetic acid-acetate buffer   50 ml
     test paper, pH 9.5-14.0
       (Baker 2880-1) or indicator
       solution (VWR 34186-839)
     controls: aniline, n-butylamine,   3 ml/1.5 g ea
       dibutylamine, p-toluidine

C-5 copper wire, 0.5 mm, 10 cm lengths 10 (2 g)
       (Baker 1736)
     controls: bromobutane,   1 ml ea
       chlorobenzene

C-6 *Benedict's reagent   75 ml
     controls: benzaldehyde,   4 ml ea
       butyraldehyde

C-7 *0.2M bromine/carbon tet.   50 ml
      carbon tetrachloride   15 ml
      controls: cyclohexene,   2 ml ea
        ethyl acetoacetate

C-8 *bromine water, saturated   200 ml
      controls: aniline, phenol   2 ml/1.5 g ea

C-9   acetone   15 ml
     *chromic anhydride-H_2SO_4 1 ml
      controls: benzaldehyde,   1 ml ea
        1-butanol, t-butyl alcohol,
        butyraldehyde

C-10 volumetric pipets, 1 ml   10
      controls: 1-bromobutane,1-chloro-   3 ml ea
      butane, chlorobenzene, toluene

C-11 *2,4-DNPH reagent   30 ml
      ethanol, 95%   20 ml
      controls: benzaldehyde,   1 ml ea
        cyclohexanone

C-12 3M aqueous ammonia   20 ml
      carbon tetrachlordie   7.5 ml
     *chlorine water   1 ml
      Dri-Na (Baker 9413)   7.5 g
      1M nitric acid   2 ml(0.13 ml conc.
                                                    HNO_3)
     *PNB reagent   15 ml
      0.3M silver nitrate   1.5 ml
                                                    (77 mg $AgNO_3$)
      sodium bicarbonate   2 g
      3M sulfuric acid   2 ml (0.33 ml
                                                    conc. H_2SO_4)
      controls: acetamide,   0.2 g/1.5 ml ea
        bromobenzene

C-13 *2.5% ferric chloride   2 ml
      controls: phenol, salicylic acid   0.6 g ea

C-14 ethanol, 95%   20 ml
     *2.5% ferric chloride   5 ml
      1M hydrochloric acid   45 ml
     *0.5M hydroxylamine HCl/ethanol   15 ml
      6M sodium hydroxide   3 ml (0.72 g NaOH)
      control: butyl acetate   1.5 ml

C-15 6M hydrochloric acid   50 ml
                                                    (25 ml conc. HCl)
      3M sodium hydroxide   75 ml (9.0 g NaOH)
      p-toluenesulfonyl chloride   5 g
        (T3,595-5)
      controls: aniline,   1.5 ml ea
        N,N-dimethylaniline,
        N-methylaniline

C-16 *0.5M iodine-KI reagent   100 ml

```
          3M sodium hydroxide                    20 ml (2.4 g NaOH)
          controls: 2-butanone, 2-propanol       2 ml ea

C-17  *Lucas' reagent                            30 ml
          controls: 1-butanol, 2-butanol,        3 ml ea
          t-butyl alcohol

C-18  ≈0.1M NaOH (standardized)                 400 ml
          ethanol, 95%                          400 ml
          *phenolphthalein indicator sol.         2 ml
          control: adipic acid                    3 g

C-19  ethanol, 95%                              15 ml
          0.1M potassium permanganate           40 ml (0.63 g KMnO₄)
          controls: 1-butanol, cyclohexene        1 ml ea

C-20  ethanol, 95%                              50 ml
          *2.5% quinhydrone/methanol (w/v)        3 ml
          controls: aniline, butylamine,          1 ml ea
          dibutylamine, N,N-dimethylaniline,
          N-methylaniline, tributylamine

C-21  1M nitric acid                             2 ml (0.13 ml
                                                    conc. HNO₃)
          *0.1M silver nitrate/ethanol           30 ml
          controls: bromobenzene, n-butyl         1 ml ea
          bromide, n-butyl chloride,
          t-butyl chloride

C-22  acetone                                     5 ml
          *sodium iodide/acetone reagent         15 ml
          controls: bromobenzene, n-butyl       1.5 ml ea
          bromide, n-butyl chloride,
          t-butyl chloride, 1,2-dichloro-
          ethane

C-23  2M aq. ammonia                             10 ml (1.35 ml
                                                    conc. NH₃)
          1M nitric acid                         50 ml (3.1 ml
                                                    conc. HNO₃)
          0.3M silver nitrate (brn. bottle)      30 ml (1.53 g AgNO₃)
          3M sodium hydroxide                     1 ml (0.12 g NaOH)
          control: benzaldehyde                   1 ml

Chemicals per 10 derivative preparations

D-1   3,5-dinitrobenzoyl chloride or              3 g
          p-nitrobenzoyl chloride                 3 g
          ethanol, 95%                           60 ml
          0.2M sodium carbonate                  60 ml
                                                    (1.27 g Na₂CO₃)

D-2   magnesium sulfate, anhydrous                5 g
          α-naphthyl isocyanate (Baker R714)      4 ml
                    or
          phenyl isocyanate (18,535-3)            4 ml
          petroleum ether (60-90°)               75 ml

D-3   *2,4-dinitrophenylhydrazine reag.        120 ml
          ethanol, 95%                          400 ml
```

```
                 ethyl acetate                          25 ml
D-4    ethanol, 95%                                      90 ml
       semicarbazide hydrochloride                        3 g
       sodium acetate (hydrated)                           5 g

D-5    ethanol, 95%                                      90 ml
       hydroxylamine hydrochloride                         3 g
       sodium acetate (hydrated)                           5 g

D-6    3M hydrochloric acid                             160 ml
                                                        (40 ml conc. HCl)
       6M hydrochloric acid                              75 ml (37.5 ml
                                                          conc. HCl)
       3M sodium hydroxide                              150 ml (18 g NaOH)
```

(Reagents for preparation of D-19 and D-15 derivatives not included. Students can dilute 3M HCl for use in gas traps.)

```
D-7    acetic acid, glacial                              75 ml
       ethanol, 95%                                      40 ml
       dioxane                                           40 ml
       xanthydrol (Baker X503)                            6 g

D-8    benzoyl chloride (B1,269-5)                       12 ml
       ethanol, 95%                                      50 ml
       3M hydrochloric acid                              30 ml (7.5 ml
                                                          conc. HCl)
       3M sodium hydroxide (10,813-8)                    30 ml (3.6 g NaOH)

D-9    benzenesulfonyl chloride                           6 ml
                 or
       p-toluenesulfonyl chloride                         9 g
       ethanol, 95%                                      50 ml
       6M hydrochloric acid                              50 ml (25 ml
                                                          conc. HCl)
       3M sodium hydroxide                              150 ml (18 g NaOH)

D-10   ethanol, 95%                                     150 ml
       α-naphthylisothiocyanate (N452-5)                  4.5 g
                 or
       phenylisothiocyanate (13,974-2)                    3 ml
       petroleum ether                                   40 ml

D-11   ethanol, 95%                                      40 ml
       ethyl acetate                                     40 ml
       methyl iodide (I850-7)                             3 ml

D-12   ethanol, 95%                                     120 ml
      *picric acid/ethanol, satd.                        75 ml

D-13   aq. ammonia, conc.                                75 ml
       ethanol, 95%                                     150 ml
       thionyl chloride (15,780-5)                       75 ml

D-14   aniline or                                        10 ml
       p-toluidine                                       12 g
       ethanol, 95%                                      60 ml
       ethyl ether                                      180 ml
       1.5M hydrochloric acid                            75 ml (9.4 ml
                                                          conc. HCl)
```

```
         thionyl chloride                        30 ml
D-15 ethanol, 95%                               220 ml
     1.5M hydrochloric acid                       2 ml (0.25 ml
                                                    conc. HCl)
     p-nitrobenzyl chloride (14,011-2)           4.5 g
     5% sodium carbonate                          40 ml
                                                    (2.1 g Na_2CO_3)
     3M sodium hydroxide                          10 ml (1.2 g NaOH
    *phenolphthalein indicator solution           1 ml

D-16 ethanol, 95%                               300 ml
     ethyl ether                                600 ml
     hydrochloric acid, conc.                   325 ml
     potassium carbonate                         50 g
     6M sodium hydroxide                        650 ml (156 g NaOH

D-17 ammonium chloride                            1.5 g
     benzylamine (B1,630-5)                       45 ml
     ethanol, 95%                                150 ml
     ethyl acetate                              150 ml
     3M hydrochloric acid                         1 ml (0.25 ml
                                                    conc. HCl)
     ligroin or pet. ether                        50 ml
     methanol                                     35 ml
     sodium methoxide                              2 g

D-18 3,5-dinitrobenzoic acid (12,125-8)          12 g
     ethanol, 95%                                200 ml
     ethyl ether, anhydrous                     300 ml
     0.5M sodium carbonate                      300 ml (15.9 g
                                                    Na_2CO_3)
     sulfuric acid, conc.                         1 ml

D-19 ethanol, 95%                                90 ml
     ethylene glycol                             30 ml
    *picric acid/ethanol, satd.                  30 ml
     thiourea (T3,355-3)                          3 g

D-20 ethanol, 95%                               100 ml
     nitric acid, conc.                          30 ml
                or
     fuming nitric acid (Baker 9628)             30 ml
     sulfuric acid, conc.                        30 ml

D-21 ethanol, 95%                                75 ml
     potassium permanganate                      35 g
     sodium bisulfite                            25 g
     6M sodium hydroxide                         12 ml (2.9 g NaOH
     6M sulfuric acid                            18 ml (6 ml
                                                    conc. H_2SO_4)
D-22 chloroacetic acid                           10 g
     ethyl ether                                300 ml
     6M hydrochloric acid                        90 ml (45 ml
                                                    conc. HCl)
     0.5M sodium carbonate                      150 ml (7.95 g
                                                    Na_2CO_3)
```

```
            8M sodium hydroxide                      45 ml (14.4 g
                                                     NaOH)

D-23   acetone                                       30 ml
       bromine                                       15 ml
       dioxane                                       30 ml
       ethanol                                      150 ml
       potassium bromide                             70 g
       1M sodium bisulfite                           30 ml (3.1 g
                                                     NaHSO₃)

D-24   α-naphthyl isocyanate (Baker R714)            4 ml
          or
       phenyl isocyanate (18,535-8)                  4 ml
       petroleum ether (60-90°)                      75 ml
       pyridine                                       8 ml
       0.5M sulfuric acid                             8 ml (0.22 ml
                                                     conc. H₂SO₄)
       triethylamine                                0.1 ml
```

Solution preparations (for 100 tests or 10 derivative preparations unless otherwise indicated):

Acetic acid-acetate buffer (C-4). Dissolve 240 g of anhydrous sodium acetate and 30 ml of acetic acid in water and make up to 500 ml.

Benedict's reagent (C-6). Dissolve 130 g of sodium citrate and 75 g of anhydrous sodium carbonate in 650 ml of water with heating; slowly add a solution of 13 g copper(II) sulfate pentahydrate in 100 ml of water, with stirring. Filter if necessary.

0.2M Bromine/carbon tetrachloride (C-7). Dissolve 16 g (about 5 ml) of bromine in carbon tetrachloride and make up to 500 ml with carbon tetrachloride.

Bromine water, saturated (C-8). Thoroughly mix 25 ml of bromine with 2.0 ℓ of water, let stand overnight or longer, and decant from any undissolved bromine.

Chlorine water (C-12). Acidify 10 ml of Clorox (or similar laundry bleach) to litmus with 6M HCl; keep in a dark bottle (solution deteriorates on standing). Alternatively, have students acidify just before use.

Chromic anhydride-sulfuric acid reagent (C-9). Mix 2.5 g of chromic anhydride (chromium trioxide) with 2.5 ml of conc. sulfuric acid and pour the suspension slowly, with stirring, into 7.5 ml of water. Cool to room temperature before using.

2,4-Dinitrophenylhydrazine reagent (C-11, D-3). (25 derivatives or 100 tests) Dissolve 9.0 g 2,4-dinitrophenylhydrazine (Aldrich D19,930-3) in 45 ml of conc. sulfuric acid.

102

Add the solution slowly, with constant stirring, to 210 ml of 95% ethanol combined with 60 ml water. Mix thoroughly and filter if necessary.

2.5% Ferric chloride (C-13, C-14). (100 phenol tests or 40 ester tests) Dissolve 0.50 g of anhydrous ferric chloride (or 0.83 g of the hexahydrate) in 20 ml of water and filter the solution.

0.5M Hydroxylamine hydrochloride/ethanol (C-14). Dissolve 5.2 g hydroxylamine hydrochloride in ethanol and make up to 150 ml with additional ethanol.

0.5M Iodine-potassium iodide reagent (C-16). Dissolve 125 g of iodine and 250 g of potassium iodide in water, make up to 1 liter.

Lucas' reagent (C-17). Dissolve 240 g of anhydrous zinc chloride (Aldrich 20,808-6) in 150 ml of concentrated hydrochloric acid, with cooling.

PNB reagent (C-12). Dissolve 1.5 g of p-nitrobenzaldehyde (Aldrich 13,017-6) in 150 ml of dimethyl sulfoxide. Keep in a brown bottle, discard if the yellow color of the reagent darkens.

Phenolphthalein indicator solution(C-18, D-15). (200 ml) Dissolve 0.10 g of phenolphthalein in 100 ml water +100 ml ethanol.

Picric acid/ethanol, saturated solution (D-12, D-19). (10 amine or 25 halide derivatives) Thoroughly mix 6 g of picric acid with 70 ml of 95% ethanol, filter if necessary.

20% Potassium hydroxide/glycerine (C-2). Dissolve 50 g of potassium hydroxide pellets (85%) in 200 ml of glycerine.

2.5% Quinhydrone/methanol (C-20). Dissolve 0.75 g of quinhydrone in 30 ml of methanol.

0.1M Silver nitrate/ethanol (C-21). Dissolve 5.1 g silver nitrate in ethanol and make up to 300 ml with additional ethanol. Store in a brown bottle.

Sodium iodide/acetone reagent (C-22). Dissolve 20 g of sodium iodide in acetone (or 2-butanone) and make up to 150 ml with the solvent. Keep in a dark bottle, discard when a definite red-brown color develops.

Topics for Report:

1. General equations are given in the text for most of these reactions. Students should rewrite them to apply to their specific unknowns.

2. (a) Aromatic aldehyde. It gives a false positive Lucas test because it is insoluble in the reagent. (b) Phenol. Phenols are oxidized by permanganate and react with bromine in organic solvents as well as water. (c) Aromatic secondary amine. (d) Amide of a low-M.W. amine and a solid carboxylic acid. (e) Benzylic halide (or allylic halide with an aromatic portion), presumably a bromide. Only benzylic and allylic compounds react rapidly with both ethanolic silver nitrate and sodium iodide in acetone. (f) Primary alcohol. The only 1^0 alcohol that would give a positive iodoform test is ethanol.

3. (a) S_a; (b) X; (c) B; (d) A_2; (e) N; (f) A_1; (g) X;

(h) S_b; (i) S_n; (j) N

4. (a) $PhCONHC_2H_5 + OH^- \rightarrow Ph\overset{\overset{\displaystyle O^{\ominus}}{|}}{\underset{\underset{\displaystyle NHC_2H_5}{|}}{C}}-OH \rightarrow PhCOOH + C_2H_5NH^- \rightarrow$

$PhCOO^- + C_2H_5NH_2$

(b) $CH_3\!\!\bigcirc\!\!OH + Br_2 \rightarrow CH_3\overset{\oplus}{\underset{\underset{\displaystyle Br}{|}\,H}{\bigcirc}}\!\!-OH + Br^- \rightarrow CH_3\!\!\bigcirc\!\!\underset{\displaystyle Br}{-}OH + HBr$ (repeat)

Attack may also be on the anion in aqueous solution

(c) See answer for Topic 2, Experiment 20.

(d)

$CH_3COOC_2H_5 + NH_2OH \rightarrow CH_3\overset{\overset{\displaystyle O^{\ominus}}{|}}{\underset{\underset{\displaystyle OC_2H_5}{|}}{\overset{\oplus}{C}-NH_2OH}} \rightarrow CH_3\overset{\overset{\displaystyle O}{\|}}{C}NHOH + C_2H_5OH$

(e) $ArSO_2Cl + Et_2NH \longrightarrow \left| Ar\overset{\overset{\displaystyle O^{\ominus}}{\|}}{\underset{\underset{\displaystyle O}{\|}}{S}}\overset{\displaystyle Cl}{\underset{\displaystyle \overset{\oplus}{N}HEt_2}{}} \right| \xrightarrow{OH^-} ArSO_2NEt_2 + H_2O + Cl^-$

(g) $\underline{t}\text{-BuOH} \xrightarrow{HCl} \underline{t}\text{-Bu}\overset{\oplus}{OH_2} \xrightarrow{-H_2O} \underline{t}\text{-Bu}^{\oplus} \xrightarrow{Cl^-} \underline{t}\text{-BuCl}$

(h) $CH_3CH_2\overset{\overset{\displaystyle Cl}{|}}{\underset{\underset{\displaystyle CH_3}{|}}{C}}-CH_3 \xrightarrow{Ag^+} CH_3CH_2\overset{\oplus}{\underset{\underset{\displaystyle CH_3}{|}}{C}}-CH_3 \xrightarrow{NO_3^-} CH_3CH_2\overset{\overset{\displaystyle ONO_2}{|}}{\underset{\underset{\displaystyle CH_3}{|}}{C}}-CH_3$

$+ \underline{AgCl}$

104

(i)
$$CH_3CH_2CH_2CH_2Br + NaI \longrightarrow \left[\overset{\delta-}{I} \text{---} \underset{\underset{CH_3CH_2CH_2}{|}}{\overset{\delta-}{CH_2}} \text{--} Br \right] \rightarrow CH_3CH_2CH_2CH_2I + NaBr$$

(j)
$$ArCOCl + CH_3CH_2CH_2CH_2OH \rightarrow Ar\underset{\underset{Cl^{\oplus}}{|}}{\overset{\overset{O^{\ominus}H}{|}}{C}} \text{-} OBu \rightarrow Ar\overset{\overset{O}{||}}{C}OBu + HCl$$

(k)
$$Et_3N + CH_3I \longrightarrow \left[\overset{\delta+}{Et_3N} \text{---} \overset{\delta-}{CH_3} \text{---} I \right] \rightarrow Et_3NCH_3^+I^-$$

(l)
$$ArOH \xrightarrow{OH^-} ArO^- \xrightarrow{ClCH_2COOH} \left[\overset{\delta-}{ArO} \text{---} \underset{\underset{COOH}{|}}{\overset{\delta-}{CH_2}} \text{--} Cl \right] \rightarrow ArOCH_2COOH + Cl^-$$

PART IV

ADVANCED PROJECTS

Project 1

If you wish to omit either objective 1 or 2 (or both),
juglone can be purchased from Aldrich Chemical Co., cat.
no. H4,700-3.

Project 2

The project can be modified by omitting objectives 1 and 2
and providing an essential oil for separation and analysis
of the components. Essential oils can be obtained from
Fritzche D & O and from Henry H. Ottens, whose addresses
are given in the comments under Experiment 25.

Project 3

This project may, of course, be modified to use domestic
plants like spinach, but the combination of botany and
chemistry makes the use of wild plants more interesting.

Project 4

p-Hydroxybenzaldehyde can be obtained from Aldrich, cat.
no. 14,408-8.

Project 5

4-t-Butylcyclohexanone can be obtained from Aldrich
(B9,230-3); the reducing agent, in toluene, can be obtained
from Aldrich (19,619-3, under "Red-al") and Fluka (71495).
Aldrich also provides an isomer mixture of the corresponding alcohol (B9,200-1) which could be compared with the
isomer mixture obtained by Vitride reduction.

Project 6

Benzoyl chloride (B1,269-5), o-toluoyl chloride (12,201-7),
m-toluoyl chloride (12,225-4) and other appropriate starting materials can be obtained from Aldrich and combined

with a variety of aromatic compounds to yield different aryl phenyl ketones. Alternatively, the first objective can be omitted and the ketones purchased directly from Aldrich, Baker, or another supplier.

Project 7

Cyclopentanone (C11,240-2), cyclohexanone (C10,218-0), cycloheptanone (C9,900-0) and various benzylic chlorides and bromides can be obtained from Aldrich Chemical Co.

Project 8

Cyclohexanol can be obtained from Aldrich, cat. no. 10,589-9; alternatively, you can begin with cyclohexanone instead. One possible synthetic route would involve a Grignard reaction between cyclohexanone and methylmagnesium bromide (or iodide) followed by dehydration to 1-methylcyclohexene, allylic bromination with NBS, and hydrolysis of the resulting mixture of bromides. An authentic sample of 3-methyl-2-cyclohexene-1-one can be obtained from Aldrich (cat. no. 19,771-8) for comparison, if desired.

Project 9

α-Pinene (14,752-4) and 18-Crown-6 (18,665-1) can be obtained from Aldrich Chemical Co.

EXPERIMENT 11.   2-Methyl-3-butyn-2-ol (IR)

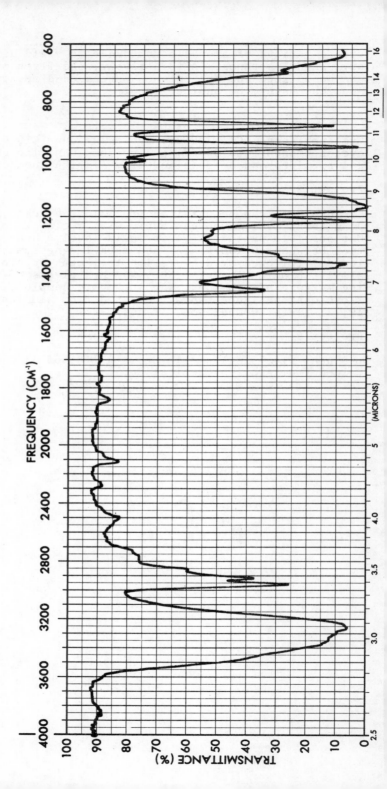

EXPERIMENT 11. 3-Hydroxy-3-methyl-2-butanone (IR)

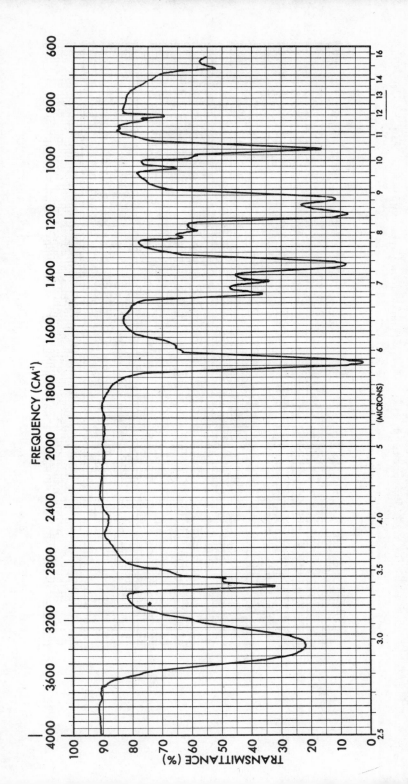

EXPERIMENT 13.   Tropylium Fluoborate (NMR)

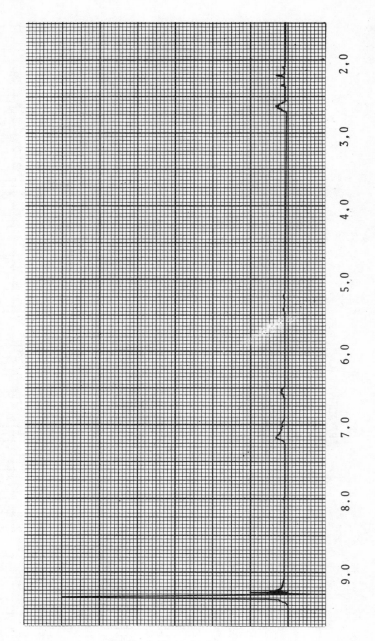

δ, ppm

EXPERIMENT 15. Lycopene-Neolycopene (UV-VIS)

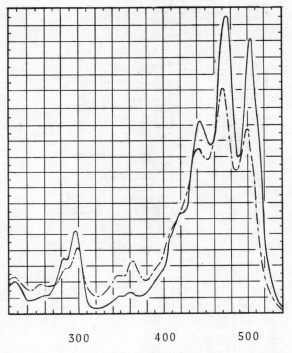

300   400   500

————Lycopene— - —— -Lycopene-Neolycopene equilibrium mixture

EXPERIMENT 22. Clove Oil

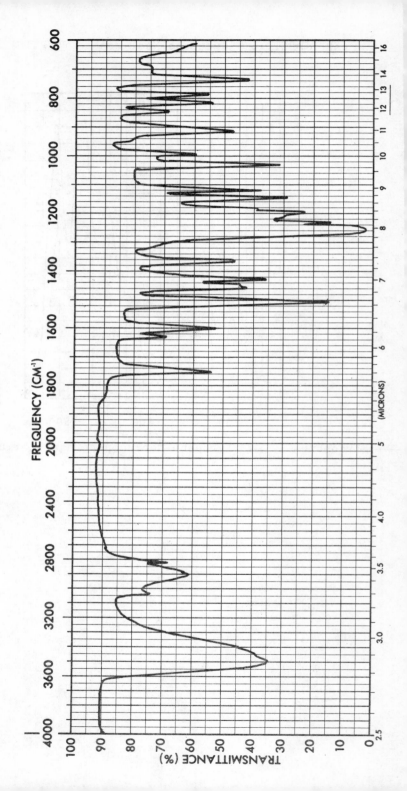

EXPERIMENT 29. "Cicelene" (IR)

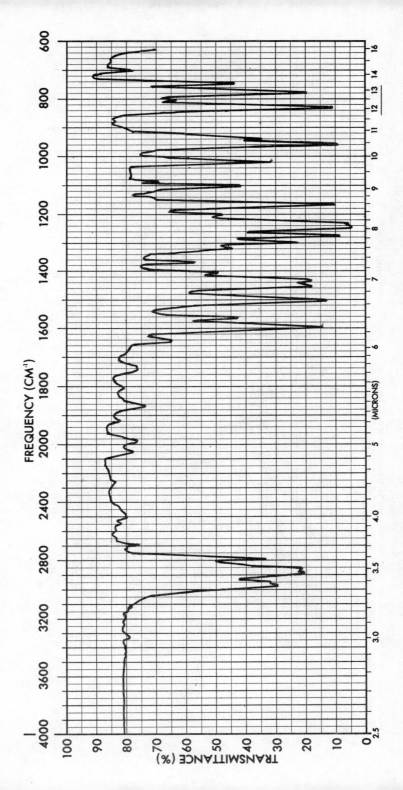

EXPERIMENT 35.   By-product spectrum (NMR)

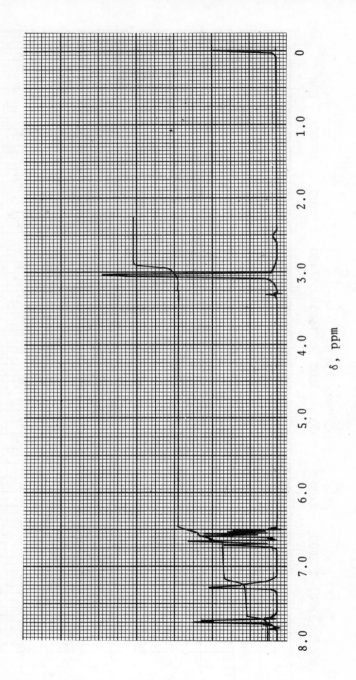

δ, ppm

EXPERIMENT 40. Dimedone (NMR)

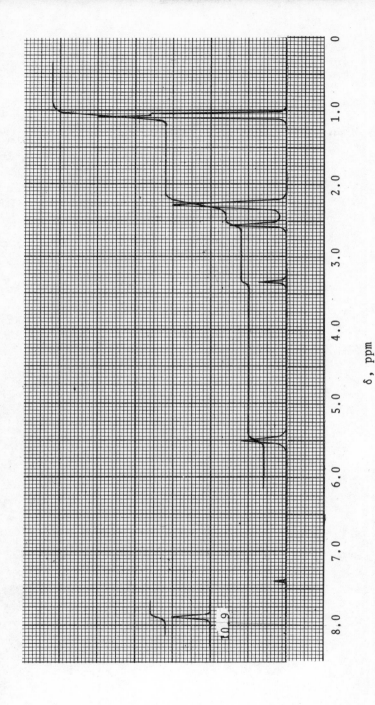

δ, ppm